T0212225

Algorithms and Software for Predictive and Perceptual Modeling of Speech

Synthesis Lectures on Algorithms and Software Engineering

Editor
Andreas Spanias, *Arizona State University*

Algorithms and Software for Predictive and Perceptual Modeling of Speech
Venkatraman Atti

ISBN: 978-3-031-00388-2   paperback
ISBN: 978-3-031-01516-8   ebook

DOI 10.1007/978-3-031-01516-8

A Publication in the Springer series
SYNTHESIS LECTURES ON ALGORITHMS AND SOFTWARE ENGINEERING

Lecture #7
Series Editor: Andreas Spanias, *Arizona State University*
Series ISSN
Synthesis Lectures on Algorithms and Software Engineering
Print 1938-1727   Electronic 1938-1735

Algorithms and Software for Predictive and Perceptual Modeling of Speech

Venkatraman Atti

Verance Corporation

SYNTHESIS LECTURES ON ALGORITHMS AND SOFTWARE ENGINEERING
#7

ABSTRACT

From the early pulse code modulation-based coders to some of the recent multi-rate wideband speech coding standards, the area of speech coding made several significant strides with an objective to attain high quality of speech at the lowest possible bit rate. This book presents some of the recent advances in linear prediction (LP)-based speech analysis that employ perceptual models for narrow- and wide-band speech coding.

The LP analysis-synthesis framework has been successful for speech coding because it fits well the source-system paradigm for speech synthesis. Limitations associated with the conventional LP have been studied extensively, and several extensions to LP-based analysis-synthesis have been proposed, e.g., the discrete all-pole modeling, the perceptual LP, the warped LP, the LP with modified filter structures, the IIR-based pure LP, all-pole modeling using the weighted-sum of LSP polynomials, the LP for low frequency emphasis, and the cascade-form LP. These extensions can be classified as algorithms that either attempt to improve the LP spectral envelope fitting performance or embed perceptual models in the LP. The first half of the book reviews some of the recent developments in predictive modeling of speech with the help of Matlab™ Simulation examples.

Advantages of integrating perceptual models in low bit rate speech coding depend on the accuracy of these models to mimic the human performance and, more importantly, on the achievable "coding gains" and "computational overhead" associated with these physiological models. Methods that exploit the masking properties of the human ear in speech coding standards, even today, are largely based on concepts introduced by Schroeder and Atal in 1979. For example, a simple approach employed in speech coding standards is to use a perceptual weighting filter to shape the quantization noise according to the masking properties of the human ear. The second half of the book reviews some of the recent developments in perceptual modeling of speech (e.g., masking threshold, psychoacoustic models, auditory excitation pattern, and loudness) with the help of Matlab™ simulations. Supplementary material including Matlab™ programs and simulation examples presented in this book can also be accessed at www.morganclaypool.com/page/atti

KEYWORDS

Linear Prediction (LP), speech coding standards, perceptual LP, warped LP, cascade-form LP, predictive modeling, LP parameter transformation, psychoacoustic model, masking asymmetry, auditory excitation pattern, and perceptual loudness

Contents

Preface

The purpose of this book is to provide an in-depth treatment of predictive and perceptual modeling techniques for wideband speech. The motivation for the proposed book came from the fact that the existing speech coding textbooks primarily focus on the theoretical aspects while limiting to high level descriptions. In this lecture series, we attempt to bridge this gap by complementing the speech coding theory taught in the classroom with a suite of Matlab™-based software modules for implementing various predictive and perceptual modeling techniques. The intended readership of this book includes at least three groups. At the highest level, any reader with a general scientific background will be able to gain an appreciation for the success of linear predictive coding in cellular telephony and psychoacoustic models in MP3 players. Secondly, readers with a general electrical and computer engineering background will become familiar with the essential signal processing techniques in speech codecs. Finally, undergraduate and graduate students with focus in DSP and multimedia will gain important knowledge in perceptual modeling and code excited linear predictive (CELP) coding algorithms. The Matlab™ programs (e.g., the psychoacoustic model, LP analysis-synthesis) are self-contained and computer exercises can be assigned in undergraduate DSP class projects.

In Chapter 1, we present an overview of the CELP-analysis synthesis. A review of various narrow and wideband speech coding standards is also given in Chapter 1. Chapter 2 focuses on the predictive modeling of speech, and Chapter 3 reviews some of the perceptual modeling techniques of speech.

This book has originated from my research at Arizona State University. Many people were instrumental in the successful completion of this book. I would like to thank Dr. Andreas Spanias and Dr. Ted Painter for their guidance during my Doctoral program at Arizona State University. I would like to thank several colleagues who had read and provided valuable feedback on the earlier versions of this book. Specifically, many thanks to Prof. Antonia Papandreou-Suppappola, Prof. Lina Karam, Prof. Tsakalis, Khawza Ahmed, Mahesh Banavar, and Visar Berisha. Finally, many thanks to Morgan & Claypool production team, Joel Claypool and Dr.C.L. Tondo for their diligent efforts in copyediting and typesetting.

Venkatraman Atti
March 2011

CHAPTER 1

Introduction

The application of perceptual models in speech coding began receiving attention during the late nineteen seventies. Methods that exploit the masking properties of the human ear in speech coding standards are largely based on concepts introduced by Schroeder and Atal in 1979 [1, 2]. In this book, we describe some of the recent advances in linear prediction (LP)-based speech analysis that employ perceptual models for narrow- and wide-band speech coding.

From the early pulse code modulation (PCM)-based speech coders to some of the recent multi-rate wideband speech coding standards, the area of *speech coding* made several significant strides with an objective to attain high quality of speech at the lowest possible bit rate. In particular, in speech coding, the objective is to attain high quality speech at low bit rates (4 to 8 kb/s), with an acceptable algorithmic delay (5 to 20 ms), and with reduced computational complexity, typically, less than 20 million instructions per second (MIPS). The speech coders can be broadly classi- fied as the following: waveform coders, transform coders, LP analysis-synthesis coders, and hybrid coders. Waveform coders (e.g., PCM, adaptive differential PCM) focus upon representing the speech waveform without necessarily exploiting the underlying speech model. Waveform coders operate at higher bit rates, e.g., 16-64 kb/s. Transform coders employ unitary transforms (e.g., Discrete Fourier Transform, Discrete Cosine Transform) for the time-frequency analysis section. Transform coders typically encode high-resolution spectral estimates at the expense of adequate temporal resolution. The correspondence between the linear prediction (LP) analysis-synthesis and the source-system speech production model has been a primary reason for its success in speech applications. Analysis- synthesis using LPC may exploit both the short-term and long-term correlation to parameterize speech. The code-excited linear prediction (CELP) coding of speech [3, 4] is proven to deliver toll-quality or near toll-quality at medium (8 to 16 kb/s) bit rates.

1.1 SPEECH ANALYSIS AND SYNTHESIS

Figure 1.1 shows a generic CELP analysis-by-synthesis architecture for speech coding. CELP synthesis relies on a source-system configuration that mimics the human synthesis system. The input speech is segmented into frames of 10 ms [5] or more typically 20 ms [6]. A pre-processing stage that consists of a high-pass filter (e.g., f_c = 140 Hz) is employed to attenuate low-frequency components in narrowband speech coding. Depending upon the application, noise suppression (NS) [7]–[10] may also be included in the pre-processing stage in order to reduce the background noise. FFT-based speech enhancement techniques have been quite popular in this regard because of their low complexity [7].

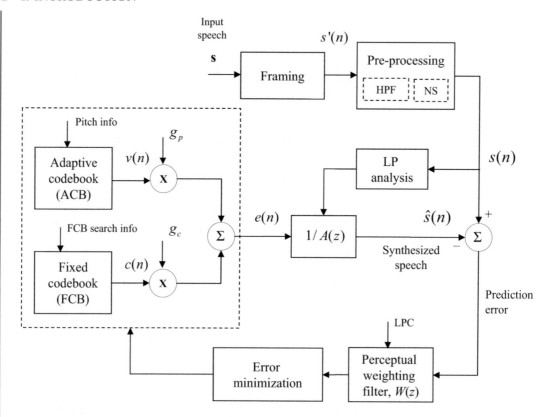

Figure 1.1: A code excited linear prediction (CELP) analysis-by-synthesis architecture for low bit rate speech coding. In the Figure, s, is the input speech, $s(n)$ is the pre-processed speech, $\hat{s}(n)$ is the reconstructed speech, $A(z)$ is the LP analysis filter, $W(z)$ is the perceptual weighting filter, $v(n)$ and $c(n)$ are the adaptive and fixed codebook vectors, g_p and g_c are the adaptive and fixed codebook gains, and $e(n)$ is the excitation, i.e., $e(n) = g_p v(n) + g_c c(n)$.

In the LP analysis stage, the prediction coefficients are estimated using the Levinson-Durbin recursion [11, 12]. The LP analysis consists of the following steps: 1) windowing of the pre-processed speech using asymmetric windows [13], 2) autocorrelation estimation from the windowed speech, and 3) bandwidth expansion of the autocorrelation sequence in order to reduce the ill-conditioning in the Levinson-Durbin recursion. The transfer function of the LP analysis filter is given by,

$$A(z) = 1 - \sum_{i=1}^{L} a_i z^{-i} \qquad (1.1)$$

where a_i are the linear prediction coefficients and L is the prediction order. Because the LP coefficients are sensitive to quantization errors, they are typically transformed to line spectrum frequency (LSF) pairs for encoding [14, 15]. The LSFs have better quantization and interpolation properties than direct-form LP coefficients [15]–[17].

The prediction residual from the LP analysis is modeled and encoded using a closed-loop analysis or analysis-by-synthesis (A-by-S) process. In A-by-S CELP systems, the excitation sequence that minimizes the "perceptually-weighted" mean-square-error (MSE) between the input and reconstructed speech is selected from a codebook [3, 4] [18]–[20]. The closed-loop LP combines the spectral modeling properties of vocoders with the waveform matching attributes of waveform coders, and hence, the A-by-S LP coders are also called *hybrid* coders. The perceptual weighting filter (PWF), $W(z)$, shapes the prediction error such that quantization noise is masked by the high-energy formants. The PWF is given by,

$$W(z) = \frac{A(z/\gamma_1)}{A(z/\gamma_2)} = \frac{1 - \sum_{i=1}^{L} \gamma_1^i a_i z^{-i}}{1 - \sum_{i=1}^{L} \gamma_2^i a_i z^{-i}} \tag{1.2}$$

where γ_1 and γ_2 are the adaptive weights and $0 < \gamma_2 < \gamma_1 < 1$. Typically, γ_1 ranges from 0.94 to 0.98, and γ_2 varies between 0.4 and 0.7, depending upon the spectral tilt or the flatness characteristics associated with the LPC spectral envelope. The PWF, $W(z)$, is adaptive, and its frequency spectrum is a smoothed version of the inverse vocal tract spectrum. Hence, the role of $W(z)$ is to de-emphasize the error energy in the formant regions [1]. This de-emphasis strategy is based on the fact that in the formant regions, quantization noise is partially masked by speech. The perceptual quantization strategies in CELP type algorithms are relatively straightforward. More elaborate perceptual models are used in the ISO/IEC MPEG standards [21]–[23] and other audio coding algorithms [24]–[26].

The pre-processed speech is passed through the perceptual weighting filter, $W(z)$, to produce the weighted speech, $s_w(n)$,

$$s_w(n) = s(n) + \sum_{i=1}^{L} a_i \gamma_1^i s(n-i) - \sum_{i=1}^{L} a_i \gamma_2^i s_w(n-i) . \tag{1.3}$$

The perceptually-weighted speech, $s_w(n)$, is used to estimate the open-loop pitch period, τ. This is because the weighted speech exhibits a better transient behavior than the pre-processed speech and, to some extent, avoids the estimation of pitch multiples. The autocorrelation of $s_w(n)$ is typically computed over the integer lags, 20 through 143 (for sampling rate, f_s=8 kHz).

$$r_{ss}(m) = \sum_{n=0}^{N-1-m} s_w(n) s_w(n+m), \quad 20 \leq m \leq 143 . \tag{1.4}$$

The lag that results in the maximum autocorrelation value, $|r_{ss}(m)|$, is selected as the initial open-loop estimate of the pitch period. This open-loop pitch estimate is further refined and a more accurate pitch period, τ_c, with *sub-sample* resolution is often computed using the closed-loop pitch analysis [5] [27, 28]. The closed-loop pitch, τ_c, is used to generate the adaptive codevector that produces the pseudo-periodic contribution in the final excitation sequence.

In CELP coding standards such as the ITU-T G.729 [5, 17] and the ITU-T G.723.1 [29], the excitation signal, $e(n)$, is generated from two codebooks (Figure 1.1), namely, the adaptive codebook (ACB) and the fixed codebook (FCB), i.e.,

$$e(n) = g_p v(n) + g_c c(n) \qquad (1.5)$$

where g_p and g_c are the ACB and FCB gains, and $v(n)$ and $c(n)$ are the ACB and FCB vectors, respectively. The adaptive and fixed codebook gains, g_p and g_c, are jointly quantized using a conjugate-structure codebook [30]. The ACB vector, $v(n)$, represents a delayed (i.e., by τ_c) segment of the past excitation signal and contributes to the periodic component of the overall excitation. In other words, high-quality speech at low bit rates can be produced by encoding non-instantaneous (i.e., delayed or past) excitation sequences in conjunction with the A-by-S optimization [2] [16, 31]. Typically, an A-by-S system consists of a short-term LP synthesis filter, $1/A(z)$, and a long-term predictor (LTP) synthesis filter, $1/A_L(z)$. A typical transfer function of a simple LTP synthesis filter, $H_\tau(z)$, is given by,

$$H_\tau(z) = \frac{1}{A_L(z)} = \frac{1}{1 - a_\tau z^{-\tau}} \qquad (1.6)$$

where τ is the pitch delay and $a_\tau = r_{ss}(\tau)/r_{ss}(0)$ is the gain parameter. However, the recent trend in speech coding standards is that the long-term correlations in speech signals are not exploited using the LTP synthesis filter explicitly; instead, a pitch prefilter in conjunction with the adaptive and fixed codebooks is a popular choice [17].

After the periodic contribution in the overall excitation is captured in $g_p v(n)$, a fixed codebook search is performed. The FCB excitation vector, $c(n)$, partly represents the remaining aperiodic component in the excitation signal and is constructed using an algebraic codebook of interleaved, unitary-pulses. The term algebraic CELP (ACELP) refers to the structure of the codebooks used to select the excitation codevector. Various algebraic codebook structures have been proposed, but the most popular is the interleaved pulse permutation code [32, 33]. In this codebook, the codevector consists of a set of interleaved permutation codes containing only a few non-zero elements (around 4 to 5 pulses are used). This is given by,

$$p_k = k + jl, \quad j = 0, 1, \ldots, 2^M - 1 \qquad (1.7)$$

where p_k is the pulse position, k is the pulse number, and l is the interleaving depth. The integer, M, is the number of bits describing the pulse positions. Table 1.1 shows an example ACELP codebook structure, where, the interleaving depth l equals 5, and the number of bits to represent the pulse positions, $M = 3$. From Eq. (1.7), $p_k = k+5j$, where $k = 0, 1, 2, 3, 4$; $j = 0, 1, 2, \ldots, 7$. For a given value of k, the set defined by Eq. (1.7) is known as track, and the value of j defines the pulse position.

From the codebook structure shown in Table 1.1, the codevector, $c(n)$, is given by,

$$c(n) = \sum_{k=0}^{4} \alpha_k \delta(n - p_k),\qquad (1.8)$$

where $\delta(n)$ is the unit impulse, α_k are the pulse amplitudes (± 1), and p_k are the pulse positions. The codevector, $c(n)$, is computed by placing the 5 pulses at the determined locations, p_k, multiplied with their signs, α_k. The pulse positions and the signs are encoded and transmitted. Note that the algebraic codebooks do not require any storage.

Table 1.1: An example algebraic codebook structure.

Track, k	Pulse positions, p_k
0	p_0: 0, 5, 10, 15, 20, 25, 30, 35
1	p_1: 1, 6, 11, 16, 21, 26, 31, 36
2	p_2: 2, 7, 12, 17, 22, 27, 32, 37
3	p_3: 3, 8, 13, 18, 23, 28, 33, 38
4	p_4: 4, 9, 14, 19, 24, 29, 34, 39

Some of the typical synthesis parameters encoded and transmitted in the analysis-by-synthesis LP include: the line spectrum pairs (LSPs), the pitch, the algebraic codevector information, and the codebook gain parameters. At the decoder, the transmitted source parameters are used to form the excitation. The excitation, $\hat{e}(n)$, is then used to excite the estimated LP synthesis filter, $1/\hat{A}(z)$, to reconstruct the speech signal. A series of adaptive post-processing filters [34] are applied on the decoded speech in order to enhance the perceptual quality. Finally, error protection and frame erasure concealment techniques are also employed to improve the performance of the speech codec under channel errors [17].

1.2   NARROWBAND AND WIDEBAND SPEECH CODERS

Signal bandwidth in narrowband LP coding is limited to 150-3400 Hz. Signal bandwidth in wideband speech coding spans from 50 Hz to 7000 Hz, which substantially improves the quality of signal reconstruction, intelligibility, and naturalness. In particular, the introduction of the low-frequency components improves the naturalness, while the higher frequency extension provides more adequate speech intelligibility. In case of high-fidelity audio, it is typical to consider sampling rates of 44.1 kHz and signal bandwidth can range from 20 Hz to 20 kHz. Some of the recent super high-fidelity audio storage formats such as the DVD-audio and the super audio CD (SACD) consider signal bandwidths up to 100 kHz [141].

Linear predictive coding is mostly used in the coding of speech signals and the dominant application of LPC is cellular telephony. Although LP modeling has been the core of compression algorithms in cellular telephony, the LP analysis-synthesis framework has also been integrated in some of the wideband speech coding standards [98] [115] and audio modeling [76], [160, 161]. Whether or not LPC is amenable for audio modeling depends on the source signal properties. For example, a code-excited linear predictive (CELP) coder is more adequate than a transform or sinusoidal coder for telephone speech, while the transform or parametric coders seem to be more promising than the CELP coder for music. Coding schemes that are multi-modal and that facilitate hybrid speech and audio coding architectures were also proposed [59, 60]. For example, the recently standardized ITU-T G.729.1 Embedded Variable bit rate (EV) wideband speech and audio codec employs such hybrid architecture [164].

The International bodies such as the European Telecommunications Standards Institute (ETSI), the Telecommunications Industry Association (TIA), and the International Telecommunications Union (ITU) develop formal speech coding standards. Table 1.2 lists various speech coding standards published by the ITU, the TIA, and the ETSI workgroups. The FS 1016 CELP [107, 108], the ITU-T G.728 LD-CELP [117], the TIA IS-54 vector-sum excited LP (VSELP) [122], and the TIA IS-96 Qualcomm CELP (QCELP) [56] belong to the first-generation CELP coding family. Second-generation CELP algorithms include the ITU-T G.723.1 dual-rate speech codec [29], the TIA IS-127 relaxed CELP (RCELP) [7], the GSM enhanced full rate (EFR) [109, 110] and the ITU-T G.729 CS-ACELP [5, 17]. The second-generation CELP algorithms are primarily targeted for use in Internet audio streaming, voice-over-Internet-protocol (VoIP), toll-quality[1] cellular telephony, and teleconferencing applications. Several methods such as the split-vector quantization of LSFs [15], fast algebraic codebook search [32], conjugate VQ encoding of the gains [30], and adaptive post-processing steps [34] contributed to the success of the second-generation speech coders. In fact, the basic infrastructure in most of the recent third-generation speech coding standards is inherited from the CS-ACELP algorithm [17], a popular second-generation toll-quality speech coding standard.

The third-generation (3G) CELP algorithms are multimodal and accommodate multiple bit rates. Some of the 3G standards include the TIA IS-893 selectable mode vocoder (SMV) [6] and the ITU-T G.722.2 adaptive multirate wideband (AMR-WB) standard [45, 115]. The SMV algorithm was developed to provide higher quality and capacity over the existing IS-96 QCELP and IS-127 EVRC codecs. Efforts to establish wideband speech coding standards [127, 128] continue to drive further the research and development towards algorithms that work at lower rates and deliver enhanced speech quality. During the early 1990s, several wideband speech coding algorithms have been proposed [156]–[159] [164]. Some of the coding principles associated with these algorithms have been successfully integrated into several wideband speech coding standards, for example, the

---

[1] Speech quality is classified into four general categories: broadcast, network/toll, communications, and synthetic. Broadcast speech refers to high quality commentary speech. Toll quality refers to quality comparable to the classical analog speech. Communications quality refers to degraded speech quality, which is nevertheless natural, highly intelligible, and adequate for telecommunications. Synthetic speech is usually intelligible but can be unnatural. The mean opinion score (MOS, a 5-level quality scale) is a measure that is widely used to quantify coded speech quality. MOS values 5, 4, 3, 2, and 1 correspond to subjective qualities of excellent, good, average, poor, and bad, respectively. A MOS value between 4 and 4.5 implies toll quality, scores between 3.5 and 4 imply communications quality, and a MOS between 2.5 and 3.5 implies synthetic quality.

Table 1.2: A list of speech coding standards. The complexity values listed in the table are system-dependent. The codec delay is due to framing and look-ahead.

Speech Standard	Average bit rate (kb/s)	Computational complexity (MIPS)	Mean opinion score (MOS)	Codec Delay (ms)	Related references
Federal Standards (FS)					
LPC10e (FS1015)	2.4	5-7	2.3	22.5	[104]–[106]
CELP (FS1016)	4.8	16	3.2	37.5	[107, 108]
European Telecommunications Standards Institute (ETSI)					
GSM Full-rate	13	4.5	< 3.9	20	[109]
GSM Half-rate	5.6	30	< 3.9	25	[111]
International Telecommunications Union (ITU) standards					
G.711 (PCM)	64	< 1	4.4	0.125	[113]
G.721 (ADPCM)	32	1.25	4.1	0.125	[114]
G.722 (Subband)	48/56/64	10	4.2	1.5	[98, 99]
G.722.2 (AMR-WB)	6.3-23.85	38.9	4.1	20	[45, 115]
G.723.1 (Dual-rate)	5.3/6.3	15-20	3.5/3.98	37.5	[29]
G.728 (LD-CELP)	16	10-15	4	0.625	[117, 118]
G.729 (CS-ACELP)	8	20	4.1	15	[5, 17]
G.729.1	8-32	35.79	4.1	48.9375	[164]
ITU-T 4 kb/s	4	20-30	4.1	40	[119]–[121]
Telecommunications Industry Association (TIA) standards					
IS-54 (VSELP)	7.95	13.5	3.5	25	[122, 123]
IS-96 (QCELP)	1.2/2.4/4.8/9.	15	3.33	25	[56]–[58]
IS-127	1.2/2.4/9.6	20	3.8	25	[7]
IS-893 (SMV)	1.2/2.4/4.8/9.	30	4.1	30	[6, 124]

ITU-T G.722 subband ADPCM standard [98, 99], the ITU-T G.722.2 AMR-WB codec [45, 115], and the ITU-T G.729.1 Embedded Variable bitrate (EV) wideband speech codec [164].

The ITU-T G.722 standard uses a combination of both subband and adaptive differential PCM (ADPCM) techniques [98, 99]. The input signal is sampled at 16 kHz and decomposed into two subbands of equal bandwidth using quadrature mirror filter (QMF) banks. The low-frequency subband is typically quantized at 48 kb/s while the high-frequency subband is coded at 16 kb/s. The G.722 coder includes an adaptive bit allocation scheme and an auxiliary data channel. Moreover, provisions for quantizing the low-frequency subband at 32 or at 40 kb/s are available. In particular, the G.722 algorithm is multimodal and can operate in three different modes, i.e., 48, 56, and 64 kb/s by varying the bits used to represent the lower band signal.

The ITU-T G.722.2 [45, 115] is an adaptive multi-rate wideband (AMR-WB) codec that operates at bit rates ranging from 6.6 to 23.85 kb/s. The G.722 AMR-WB standard is primarily targeted for the voice-over IP (VoIP), 3G wireless communications, ISDN wideband telephony, and audio/video teleconferencing. The AMR-WB codec has also been adopted by the third-generation partnership project (3GPP) for GSM and the 3G WCDMA systems for wideband mobile communications [45]. This, in fact, brought to the fore all the interoperability-related advantages for wideband voice applications across wireline and wireless communications. The ITU-T G.722.2 AMR-WB codec is based on the ACELP coder and operates on audio frames of 20 ms sampled at 16 kHz. The codec supports the following nine bitrates: 23.85, 23.05, 19.85, 18.25, 15.85, 14.25, 12.65, 8.85, and 6.6 kb/s. The two lowest modes, i.e., the 8.85 kb/s and the 6.6 kb/s are intended for transmission over noisy time-varying channels. Other encoding modes, i.e., 23.85 kb/s through 12.65 kb/s, offer high quality signal reconstruction. The G.722 AMR-WB includes several innovative techniques such as, 1) a modified perceptual weighting filter that decouples the formant weighting from the spectrum tilt, 2) an enhanced closed-loop pitch search to better accommodate for the variations in the voicing level, and 3) efficient codebook structures for fast searches. The codec also includes a voice activity detection scheme that activates a comfort noise generator module (1 to 2 kb/s) in case of discontinuous transmission.

The ITU-T G.729.1 embedded variable bit rate (EV) standard [164] is a scalable wideband speech and audio codec that operates at bitrates ranging from 8 kb/s to 32 kb/s. The G.729.1EV algorithm employs a three-stage coding structure, i.e., ACELP coding for low band from 50 Hz to 4000 Hz, parametric coding based on time-domain bandwidth extension algorithm for the high band from 4000 Hz to 7000 Hz, and a predictive transform coding based on time-domain aliasing cancellation algorithm for the full band from 50 Hz to 7000 Hz. The ITU-T G.729.1 EV codec operates on audio sampled at 16 kHz and produces bitstream that is scalable and consists of 12 embedded layers. Layer 1 is the core layer corresponding to a bit rate of 8 kb/s and is compliant with the ITU-T G.729 bitstream. This bitstream compliance enables that at 8 kb/s, the G.729EV is fully interoperable with G.729. Layer 2 is a 4 kb/s narrowband enhancement layer. Layers 3 to 12 are wideband enhancement layers that add 20 kb/s in steps of 2 kb/s. The G.729.1EV standard is primarily targeted for wideband telephony and voice-over IP applications.

1.3   ORGANIZATION OF THE BOOK

This book is organized as follows. In Chapter 2, a review of various linear prediction-based speech analysis and synthesis methods is presented. In Chapter 3, we focus on how certain perceptual features from the cochlear response are processed to obtain a useful representation of the signal. Such representations include the masked threshold, the auditory excitation pattern, the masking pattern, and the loudness.

CHAPTER 2

Predictive Modeling of Speech

Voiced speech is produced by exciting the vocal tract with quasi-periodic glottal air pulses generated by the vibrating vocal chords. Unvoiced speech is produced by forcing air through a constriction in the vocal tract. Nasal sounds (e.g., /n/, /m/) are due to the acoustical coupling of the nasal tract to the vocal tract. Plosive sounds (e.g., /p/) are produced by abruptly releasing air pressure which was built up behind a closure in the tract. The linear prediction (LP) analysis-synthesis framework Figure 2.1 has been successful for speech coding because it fits well the source-system paradigm for speech synthesis [11, 48]. In particular, the slowly time-varying spectral characteristics of the upper vocal tract (i.e., system) are modeled by an all-pole filter, while the prediction residual captures the voiced, unvoiced, or mixed excitation behavior of the vocal chords (i.e., source). The LP analysis

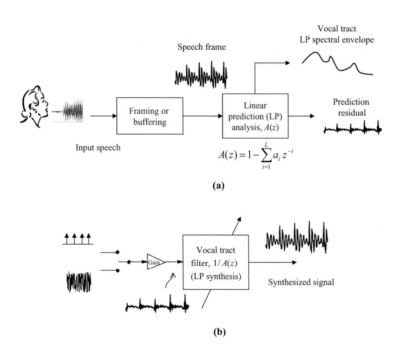

Figure 2.1: Linear prediction analysis-synthesis (a) Parameter estimation using LP analysis, (b) Speech. synthesis.

filter, $A(z)$, is given by,

$$A(z) = 1 - \sum_{i=1}^{L} a_i z^{-i} \tag{2.1}$$

where L is the order of the linear predictor. Figure 2.1 (b) depicts a simple speech synthesis model where a time-varying digital filter is excited by quasi-periodic waveforms when speech is voiced (e.g., as in steady vowels) and random waveforms for unvoiced speech (e.g., as in consonants). The inverse filter, $1/A(z)$, shown in Figure 2.1 (b), is known as the all-pole LP synthesis filter,

$$H(z) = \frac{1}{A(z)} = \frac{1}{1 - \sum_{i=1}^{L} a_i z^{-i}}. \tag{2.2}$$

The source-system model became associated with Autoregressive (AR) time-series methods where the vocal tract filter is all-pole, and its parameters are obtained by LP analysis [48]. Itakura and Saito [14, 54] and Atal and Schroeder [2] were the first to apply LP techniques to speech. Atal and Hanauer [12] later reported an analysis-synthesis system based on LP. Theoretical and practical aspects of linear predictive coding (LPC) were examined by Markel and Gray [11], and the problem of spectral analysis of speech using linear prediction was addressed by Makhoul [49].

Voiced speech is harmonically structured in the frequency-domain while unvoiced speech is random-like and broadband. The short-time spectrum of voiced speech is characterized by both fine and formant structure. The fine harmonic structure is due to the quasi-periodicity of speech and can be attributed to the vibrating vocal chords. The formant structure is due to the interaction of the source and the vocal tract. The spectral envelope is characterized by a set of peaks called formants (F1, F2, F3, and F4 as shown in Figure 2.2). The formants are the resonant modes of the vocal tract. The amplitudes and locations of the first three formants, usually occurring below 3 kHz, are more important both in speech synthesis and perception.

The classical LP-based analysis-synthesis model has been successful in cellular telephony mainly because of its simplicity and its inherent spectral whitening feature. Another supportive feature associated with the LP analysis-synthesis is the availability of several LP parameter transformations (e.g., line spectrum pairs [15] and immittance spectrum pairs [45, 62]) that enable efficient quantization and encoding of prediction parameters. These appealing features of linear predictive coding motivated several researchers to further improve the codec performance over a wide variety of source signals [39]–[42] [46, 63, 64]. Some of the extensions to the conventional LP include:

- the perceptual LP [40],

- the warped LP [41, 46],

- the discrete all-pole modeling [42] and [37, 38],

- the LP with modified filter structures [63],

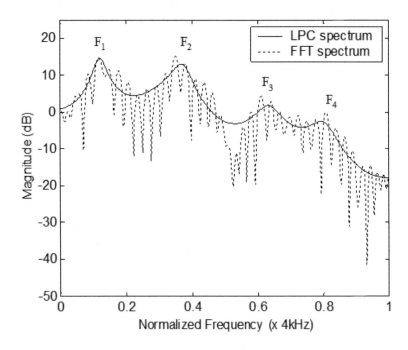

Figure 2.2: The LP and FFT spectra (dotted line). The formants represent the resonant modes of the vocal tract.

- the IIR-based pure LP [64],

- the LP for low frequency emphasis [39],

- the weighted sum of LSP polynomials [65] and,

- the cascade-form LP [66].

2.1   THE FORWARD LINEAR PREDICTION

The main idea in forward LP is to predict the current speech sample, $s(n)$, through a linear combination of previous L samples. This is given by,

$$\hat{s}(n) = a_1 s(n-1) + a_2 s(n-2) + \cdots + a_L s(n-L)$$
$$= \sum_{i=1}^{L} a_i s(n-i) \tag{2.3}$$

where a_i, $i = 1,2,... L$ are the prediction coefficients, L is the prediction order, and $\hat{s}(n)$ is the predicted speech sample. The direct-form LP analysis filter is depicted in Figure 2.3. The error, $e(n)$, that results from the prediction process is given by,

$$e(n) = s(n) - \hat{s}(n)$$

$$= s(n) - \sum_{i=1}^{L} a_i s(n - i) . \tag{2.4}$$

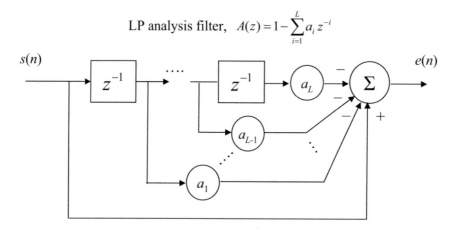

LP analysis filter, $A(z) = 1 - \sum_{i=1}^{L} a_i z^{-i}$

Figure 2.3: Linear prediction (LP) analysis. In this figure, $s(n)$ is the input speech sample, a_i, $i = 1, 2,...$ L are the prediction coefficients, L is the prediction order, and $e(n)$ is the prediction residual.

The LP coefficients, a_i, are estimated through a least-square minimization of the prediction residual, $e(n)$. The cost function to be minimized is given by $\varepsilon = \sum_n \{e^2(n)\}$

$$\frac{\partial \varepsilon}{\partial a_i} = 0, \quad 1 \leq i \leq L \tag{2.5}$$

$$\frac{\partial}{\partial a_i} \left[\sum_n \left\{ s(n) - \sum_{k=1}^{L} a_k s(n - k) \right\}^2 \right] = 0$$

$$2 \sum_n \left\{ s(n) - \sum_{k=1}^{L} a_k s(n - k) \right\} s(n - i) = 0 .$$

Rearranging the terms in the above equation,

$$\sum_{k=1}^{L} a_k \sum_n s(n - k)s(n - i) = \sum_n s(n)s(n - i), \quad 1 \leq i \leq L . \tag{2.6}$$

Depending upon the choice of the interval, n, the normal equations in Eq. (2.6), can be solved using either the autocorrelation method or the covariance method [48]. For example, in the autocorrelation method, the L points outside the interval are not used. In other words, a window is applied to $s(n)$ such that all points outside a desired finite interval are zero.

$$s_w(n) = \begin{cases} s(n)w(n), & n = 0, 1, 2, \ldots N - 1 \\ 0, & \text{else} \end{cases} \tag{2.7}$$

where $s_w(n)$ is the windowed signal, $w(n)$ is the LP analysis window, and N is the window length. The LP analysis window is typically 2 to 3 times the average pitch period [11]. Typical window types used in LP analysis include the rectangular window, the Hamming window as well as asymmetric windows [17, 67]. Equation (2.6) can be written as

$$\sum_{k=1}^{K} a_k r_{ss}(i - k) = r_{ss}(i), \quad 1 \le i \le L , \tag{2.8}$$

where $r_{ss}(i)$ is the autocorrelation estimate of the signal, $s_w(n)$,

$$r_{ss}(i) = \sum_{n=0}^{N-1-i} s_w(n)s_w(n + i), \quad i \ge 0 ,$$

Eq. (2.8) can be written in matrix form as,

$$\begin{bmatrix} a_1 \\ a_2 \\ a_3 \\ \vdots \\ a_L \end{bmatrix} = \begin{bmatrix} r_{ss}(0) & r_{ss}(-1) & r_{ss}(-2) & \cdots & r_{ss}(1-L) \\ r_{ss}(1) & r_{ss}(0) & r_{ss}(-1) & \cdots & r_{ss}(2-L) \\ r_{ss}(2) & r_{ss}(1) & r_{ss}(0) & \cdots & r_{ss}(3-L) \\ \vdots & \vdots & \vdots & \vdots & \vdots \\ r_{ss}(L-1) & r_{ss}(L-2) & r_{ss}(L-3) & \cdots & r_{ss}(0) \end{bmatrix}^{-1} \begin{bmatrix} r_{ss}(1) \\ r_{ss}(2) \\ r_{ss}(3) \\ \vdots \\ r_{ss}(L) \end{bmatrix} \tag{2.9}$$

or more compactly,

$$\mathbf{a} = \mathbf{R}_{ss}^{-1}\mathbf{r}_{ss} \tag{2.10}$$

where $\mathbf{a}$ is the LP coefficient vector, $\mathbf{r}_{ss}$ is the autocorrelation vector, and $\mathbf{R}_{ss}$ is the autocorrelation matrix. Note that $\mathbf{R}_{ss}^{-1}$ in Eq. (2.9) has a Toeplitz and symmetric structure. Typically, an order-recursive algorithm, such as the Levinson-Durbin recursion [11] is used to compute the LP coefficients. A Matlab implementation of the Levinson-Durbin recursion algorithm to estimate the LP coefficients is shown in Figure 2.4.

Figure 2.5 (a) shows a 20 ms voiced speech segment that is sampled at 8 kHz. Figure 2.5 (b) shows the fast Fourier transform (FFT) spectrum (dotted line) and a tenth-order LP spectral envelope (solid line) corresponding to the voiced speech segment. The time-domain and frequency-domain plots of the prediction residual obtained from the tenth-order conventional LP analysis

```
%~~~~~~~~~~~~~~~~~~~~~~~~~~~~~~~~~~~~~~~~~~~~~~~~~~~~~~~~~~~~~~~~~~~~~~~~~~~~~~~~~~
%% The Levinson-Durbin Recursion Algorithm
%~~~~~~~~~~~~~~~~~~~~~~~~~~~~~~~~~~~~~~~~~~~~~~~~~~~~~~~~~~~~~~~~~~~~~~~~~~~~~~~~~~
function lpCoeff = lpAnalysis(s, lpOrder)

  N = length(s);  % Obtain the length of speech frame
  s = s(:);    % Make sure the speech frame is a column vector

  % Compute autocorrelation vector
  s_fft = fft(s, 2*2^nextpow2(N));
  acorr = ifft(abs(s_fft).^2);
  acorr = acorr./(N-1); % Biased autocorrelation estimate

  % Levinson-Durbin recursion
  r = acorr(1 : lpOrder+1).';

  % Initialize the variables used in the LP recursion.
  A = zeros(lpOrder+1, lpOrder+1);
  delta = zeros(1, lpOrder);
  Ener(1) = r(1);
  gamma(1) = 1;   % reflection coefficient
  A(1,1) = 1;
  p = 0;
  for ic = 1 : 1 : lpOrder
    Dum = A(ic,1:ic);
    R = r(ic+1 : -1 : 2);
    sum_d = 0;
    for j = 1 : 1 : ic
      sum_d = sum_d + Dum(1, j) .* R(1, j);
    end

    delta(ic) = sum_d;
    gamma(1, ic+1) = -1 * (delta(1,ic) / Ener(1,ic));
    A(ic+1,1:ic+1) = [A(ic,1:1:ic), p] + gamma(ic+1) * [p, A(ic,ic:-1:1)];
    Ener(ic+1) = (1 - (gamma(ic+1)^2)) * Ener(ic);
  end

  lpCoeff = real(A(lpOrder+1,:));
  reflec = gamma;
```

Figure 2.4: LP analysis. The Levinson Durbin recursion algorithm.

are shown in Figure 2.5 (c) and Figure 2.5 (d), respectively. From Figure 2.5 (b), it can be noted that the all-pole LP model fits the FFT spectrum reasonably well and offers a sit-on-top effect. However, the LP fails to model well narrow and deep spectral valleys. Second, it is evident from Figure 2.5 (c)-(d) that the LP analysis filter, $A(z)$, places equal emphasis on all the frequency bins, and attempts to whiten the FFT spectrum of the input speech frame. Equal emphasis placed by the conventional LP on all frequencies, while minimizing the residual error energy, is inconsistent with the non-uniform frequency selectivity properties of the human auditory system [43, 44]. Algorithms that address specifically this inconsistency such as the perceptual LP [40] and the warped LP [41]

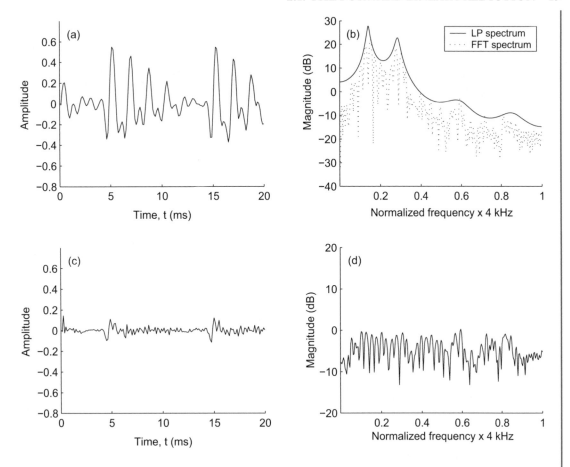

Figure 2.5: (a) A 20 ms voiced speech segment, $s(n)$, sampled at 8 kHz, (b) the FFT spectrum (dotted line) and a tenth-order LP spectral envelope (solid line) of the input speech frame, (c) the prediction residual, $e(n)$, and (d) the FFT spectrum of the prediction residual.

have been proposed. Before we describe these methods, let us focus on the choice of prediction order in conventional LP analysis.

In the above example, we chose the prediction order as $L = 10$. Typically, the prediction order depends on the sampling frequency of the input speech, and the idea is to attain a spectral resolution in the range of 400-500 Hz [12]. For example, in case of narrowband speech, $f_s = 8000$ Hz, and for a spectral resolution of $f_r = 400$ Hz (or 500 Hz), the prediction order, $L = f_s/2f_r = 10$ (or 8). In wideband speech coding ($f_s = 16$ kHz), a prediction order between 16 and 20 is typically selected. A suitable prediction order can also be calculated by measuring the performance of the predictor for various values of L. The performance of a linear predictor, for various prediction orders, can be

Table 2.1: SFM of the input speech frame shown in Figure 2.5 and the corresponding prediction residuals for various prediction orders, L. Prediction gain, p_g, values are also included for comparison.

L	SFM	p_g (dB)
0	0.2929	0
5	0.7009	6.3263
10	0.8398	9.0680
15	0.8495	10.5072
20	0.8545	11.4402
40	0.8548	12.1155
50	0.8587	12.6077

evaluated using the prediction gain,

$$p_g = 10 \log_{10} \left(\frac{\sigma_s^2}{\sigma_e^2} \right) \tag{2.11}$$

where σ_s^2 and σ_e^2 are the variances of the input signal and the prediction error, respectively. In Figure 2.6, we plotted the prediction gain of a narrowband and a wideband speech for various values of L. Considering the fact that the computational complexity of the Levinson-Durbin algorithm is of the order of L^2, it is always a challenge to choose a suitable prediction order for maximal "whitening" performance. However, several experimental studies [11, 12], [15, 16] have shown that for prediction orders greater than 20, relatively little prediction gain improvements are obtained for speech sampled at 16 kHz. Similarly, in case of narrowband speech, the prediction gain tends to saturate after $L = 10$.

Another metric that is used to study the spectral whitening performance of a linear predictor is the spectral flatness measure (SFM). The SFM of the prediction residual, $e(n)$, is defined as the ratio of geometric mean of $|E(\Omega_k)|$ to the arithmetic mean of $|E(\Omega_k)|$, i.e.,

$$SFM = \frac{\left(\prod_{k=0}^{N-1} |E(\Omega_k)| \right)^{\frac{1}{N}}}{\frac{1}{N} \sum_{k=0}^{N-1} |E(\Omega_k)|} \tag{2.12}$$

where $|E(\Omega_k)|$ is the absolute value of the DFT of prediction residual and N is the number of DFT bins. The SFM ranges between 0 and 1. An SFM of 1 implies that the residual, $e(n)$, has white noise-like characteristics. An SFM close to zero means that the signal is correlated and has a spectral tilt. In Table 2.1, we listed the SFM values of both the voiced speech segment (shown in Figure 2.5) and the prediction residuals for various prediction orders. From this table, it is evident that the SFM

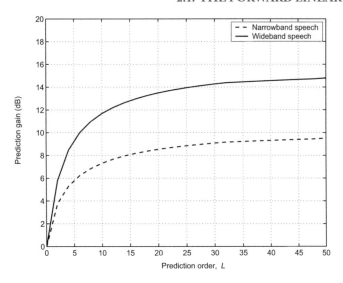

Figure 2.6: Prediction gain, p_g, versus the prediction order, L. Note that for prediction orders greater than 20, relatively little performance improvements are obtained for wideband speech. Similarly, in case of narrowband speech, the prediction gain tends to saturate for $L > 10$.

```
%~~~~~~~~~~~~~~~~~~~~~~~~~~~~~~~~~~~~~~~~~~~~~~~~~~~~~~~~~~~~~~~~~~~~~~~~~~~~~~~~~~~~~
%% Prediction gain and SFM calculation
%~~~~~~~~~~~~~~~~~~~~~~~~~~~~~~~~~~~~~~~~~~~~~~~~~~~~~~~~~~~~~~~~~~~~~~~~~~~~~~~~~~~~~
function [predGain, SFM] = predGain_SFM(s)

  N = length(s);  % Obtain the length of speech frame
  predGain = [];
  SFM = [];
  % Estimate the prediction gain and SFM for different LP orders
  for lpOrder = 0 : 2 : 50

    % LP analysis
    lpCoeff = lpAnalysis(s, lpOrder);
    prediction_error = filter(lpCoeff, 1, s);

    % Prediction gain
    var_s = sum(s.^2);
    var_e = sum(prediction_error.^2);
    predGain = [predGain, 10*log10(var_s/var_e)];

    % Spectral flatness measure
    predError_fft = fft(prediction_error, 2^nextpow2(N));
    geoMean = 10^mean(log10(abs(predError_fft)));
    arithMean = mean(abs(predError_fft));
    SFM = [SFM, geoMean/arithMean];
  end
```

Figure 2.7: Prediction gain calculation and spectral flatness measure (SFM) estimation for various prediction orders.

increases rapidly from 0.2929 (*L*=0) to 0.8398 (*L*=10) and then saturates around 0.8587 (*L*=50). Spectral whitening performance improvement for prediction orders greater than 10 is generally minimal for narrowband speech. This is coherent with the results obtained from the prediction gain experiment shown in Figure 2.6. A Matlab implementation to estimate the prediction gain and the SFM is given in Figure 2.7.

2.2   THE PERCEPTUAL LP

From the psychophysical experiments [43, 44], it is evident that the human ear resolves low-frequency sounds with relatively a higher precision. This and the physical structure of the cochlear sensors mo-tivated researchers to characterize the human auditory processing using a set of overlapping bandpass filters whose bandwidths increase as center frequencies increase (Figure 2.8). Several perceptually-significant scales that closely approximate the non-uniform frequency tiling of the auditory filter-

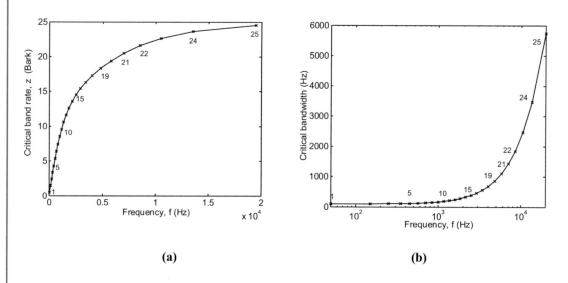

(a)   (b)

Figure 2.8: (a) Bark scale versus Hertz scale. The "*" markers denote the center frequencies of the criti-calband filters. (b) Bandwidths of the CB filter-banks. Note that at higher frequencies, the bandwidths of these filters increases.

bank have been proposed, e.g., the Mel scale [68], the Bark scale [43], and the equivalent rectangular bandwidth (ERB) scale [69]. We will defer a detailed discussion on the characterization of the au-ditory filter-bank until Chapter 3, Section 3.2.

Figure 2.10 describes the PLP analysis. First, the input speech that is sampled at f_s is passed through a bank of bandpass filters. The center frequencies of these bandpass filters are equally spaced on a Bark scale (see Figure 2.8). The Bark scale, z, is derived from the Hertz scale, f, using the

```
%~~~~~~~~~~~~~~~~~~~~~~~~~~~~~~~~~~~~~~~~~~~~~~~~~~~~~~~~~~~~~~~~~~~~~~~~~~~~~~~~~~~~~
%% Bark scale and Hz scale
%~~~~~~~~~~~~~~~~~~~~~~~~~~~~~~~~~~~~~~~~~~~~~~~~~~~~~~~~~~~~~~~~~~~~~~~~~~~~~~~~~~~~~
function [z] = Hz2Bark(f)

  % Hz to Bark scale
  fKhz = f/1000;
  z = 13 * atan(0.76*fKhz) + 3.5 * atan((fKhz/7.5).^2);

%%% Bark to Hz conversion
function [f] = Bark2Hz(z)

  % Use interpolated table lookup
  f_ic = [20 : 20000];
  z_ic = Hz2Bark(f_ic);
  f = interp1(z_ic, f_ic, z, 'spline');
```

Figure 2.9: Bark scale to Hertz scale conversion and vice-versa.

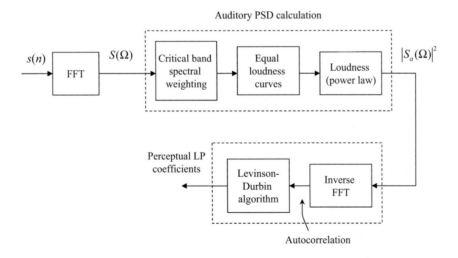

Figure 2.10: Perceptual linear prediction (PLP) analysis.In the perceptual LP (PLP) analysis [40], a perceptually-based auditory spectrum is obtained by filtering the input speech using a filter-bank that mimics the criticalband structure of the auditory filter-bank. An all-pole filter that approximates the auditory spectrum is then computed using the autocorrelation method [16].

warping function [40],

$$z = 6 \ln \left[\frac{f}{600} + \sqrt{\left(\frac{f}{600}\right)^2 + 1} \right]. \tag{2.13}$$

Usually, the above criticalband filtering operation is performed in the frequency domain. In particular, a criticalband spectral weighting function, $CB_k(\Omega)$, that emulates the filtering characteristics of the auditory filter-bank is applied on the N-point DFT spectrum, $S(\Omega)$. The criticalband spectral weighting in Figure 2.10 is given by,

$$CB_k(\Omega) = \begin{cases} 10^{(z-z_k+0.5)} & z \leq z_k - 0.5 \\ 1 & z_k - 0.5 < z \leq z_k + 0.5 \\ 10^{-2.5(z-z_k-0.5)} & z > z_k + 0.5 \end{cases} \tag{2.14}$$

where $z_k = 0.994K$ is the center frequency of the K-th criticalband in the Bark domain and $\Omega = 2\pi f / f_s$ is frequency in radians. Next, the unequal sensitivity of the human hearing at different frequencies is approximated by pre-emphasizing the output of the criticalband filter. This pre-emphasis is carried out using the equal loudness function, Figure 2.12 (d),

$$E(\Omega) = 1.151 \sqrt{\frac{\Omega^2 \left(\Omega^2 + 144 \times 10^4\right)}{\left(\Omega^2 + 16 \times 10^4\right) \left(\Omega^2 + 961 \times 10^4\right)}}. \tag{2.15}$$

The resulting auditory spectrum is converted to loudness domain as follows,

$$S_k = \sqrt[3]{E(\Omega) \int_0^\pi CB_k(\Omega) |S(\Omega)|^2 \, d\Omega}, \quad k = 0, 1, \ldots, K - 1 \tag{2.16}$$

A simple linear interpolation is performed to convert the K-point auditory spectrum, S_k, to a $(N/2+1)$-point auditory power spectrum, $|S_a(\Omega)|^2$. An N-point inverse FFT of the auditory PSD results in the autocorrelation coefficients of the speech. The PLP coefficients are then obtained using the Levinson-Durbin recursion algorithm. A Matlab implementation of the PLP analysis is given in Figure 2.11.

Figure 2.12 (a)-(f) shows the plots of intermediate signals and parameters in the PLP analysis. We use the same 20 ms voiced speech segment Figure 2.5 considered in the conventional LP example. Figure 2.12 (b)-(f) are shown on a normalized Bark scale. The PLP spectrum shown in Figure 2.12 (f) is not suitable for PLP synthesis because of the auditory approximations carried out in the analysis stage. Therefore, three post-processing steps that include amplitude unwarping, frequency unwarping, and de-emphasis are required. A triple auto-convolution of the impulse response of the PLP synthesis filter results in amplitude unwarping. Frequency scale unwarping is performed by using the all-pass filter, $(z^{-1} - \lambda)/(1 - \lambda z^{-1})$, where λ is chosen using Eq. (2.19) [69]. The de-emphasis is carried out using a two-pole lowpass filter, $1/(1 - 0.9z^{-1})^2$. The PLP analysis received

```
%~~~~~~~~~~~~~~~~~~~~~~~~~~~~~~~~~~~~~~~~~~~~~~~~~~~~~~~~~~~~~~~~~~~~~~~~~~~~~~~~~~~~~
%% PLP analysis
%~~~~~~~~~~~~~~~~~~~~~~~~~~~~~~~~~~~~~~~~~~~~~~~~~~~~~~~~~~~~~~~~~~~~~~~~~~~~~~~~~~~~~
function plpCoef = plpAnalysis(speech, fs, lpOrder)

  N = length(speech);    % Number of samples in the speech frame
  fftSize = 2^nextpow2(N);

  % Compute the power spectrum
  pspectrum = psd(speech, fftSize, fs);
  % Or you can do: pspectrum = abs(fft(speech, fftSize)).^2;

  [nfreqs,nframes] = size(pspectrum);
  nyqbar = hz2bark(fs/2);
  nfilts = ceil(nyqbar)+1;

  % Bark per filt
  step_barks = nyqbar/(nfilts - 1);

  % Bark frequency of every bin in FFT
  binbarks = hz2bark([0:(nfreqs-1)]*(fs/2)/(nfreqs-1));

  % Weights to collapse FFT bins into auditory channels
  wts = zeros(nfilts, nfreqs);
  for i = 1:nfilts
    f_bark_mid = (i-1) * step_barks;
    % Equation (2.14)
    wts(i,:) = 10.^(min(0, ...
                  min([binbarks - f_bark_mid + 0.5; ...
                      -2.5*(binbarks - f_bark_mid - 0.5)])));
  end
```

Figure 2.11: Perceptual LP (PLP) analysis program. *Continues.*

attention in spectral peak frequency estimation for vowels [40]. Later, the PLP analysis concepts were integrated in frontend analysis for feature extraction in speech recognition [71] and in relative spectral (RASTA) processing of speech [72].

2.3   THE WARPED LP

Frequency axis warping to achieve non-uniform frequency resolution was studied in [73]. The idea was later extended to warped LP by Strube [41], and was ultimately applied in an ADPCM codec [74]. Laine et al. [75] and Harma [76] used the warped LP concepts in wideband audio coding. In the warped LP [41, 46], the main idea is to warp the frequency axis, usually, according to a Bark scale prior to performing the LP analysis to effectively provide a better resolution at some frequencies. This is typically done by replacing the unit-delay elements of the LP analysis filter with all-pass sections, i.e.,

$$W_{LP}(z) = \frac{z^{-1} - \lambda}{1 - \lambda z^{-1}} , \tag{2.17}$$

```
% Compute the auditory spectrum by applying the weights on the power spectrum
aspectrum = wts * pspectrum;
[nbands,nframes] = size(aspectrum);

% bark per band
nyqbar = hz2bark(fs/2);
step_barks = nyqbar/(nbands - 1);

% Equal-loudness-curve formula, Equation (2.15)
bandcfhz = bark2hz([0:(nbands-1)]*step_barks);
fsq = bandcfhz.^2;
ftmp = fsq + 1.6e5;
eql = ((fsq./ftmp).^2) .* ((fsq + 1.44e6)./(fsq + 9.61e6));

% weight the critical bands
z = repmat(eql',1,nframes).*aspectrum;

% cube root compress, Equation (2.16)
z = z.^(.3333);

% Calculate autocorrelation
[nbands,nframes] = size(post_aspec);
r = real(ifft([post_aspec; post_aspec([(nbands-1):-1:2],:)]));

% First half only
r = r(1:nbands,:);

% Find LPC coeffs using Levinson-Durbin recursion
[plpCoef, e] = levinson(r, lpOrder);

%%% Hz to Bark conversion
function [z] = hz2bark(f)

  % Hz to Bark scale
  fKhz = f/1000;
  z = 13 * atan(0.76*fKhz) + 3.5 * atan((fKhz/7.5).^2);

%%% Bark to Hz conversion
function [f] = bark2hz(z)

  % Use interpolated table lookup
  fi = [20 : 20000];
  zi = hz2bark(fi);
  f = interp1(zi,fi,z,'spline');
```

Figure 2.11: *Continued.* Perceptual LP (PLP) analysis program.

where λ is the warping coefficient. The digital filter representation of a unit-delay element and an all-pass filter is shown in Figure 2.13.

The warping filter, $W_{LP}(z)$, passes all the frequencies un-attenuated (i.e., $\left| W_{LP}(e^{j\Omega}) \right| = 1$), hence, the name all-pass filter. The phase response of $W_{LP}(e^{j\Omega})$ is given by,

$$\angle W_{LP}(e^{j\Omega}) = \arctan\left(\frac{-\lambda + e^{-j\Omega}}{1 - \lambda e^{-j\Omega}}\right)$$

$$= \Omega + \arctan\left(\frac{\lambda \sin(\Omega)}{1 - \lambda \cos(\Omega)}\right). \tag{2.18}$$

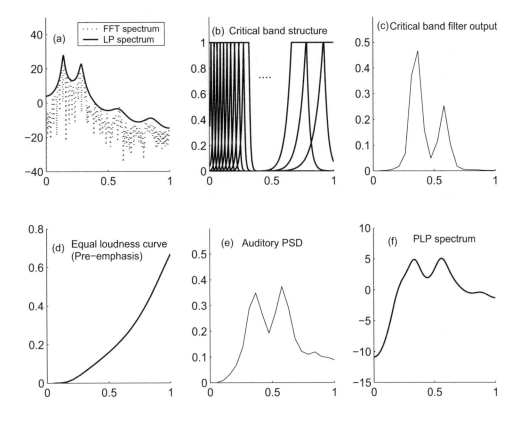

Figure 2.12: PLP analysis. Plots (b) through (f) are shown on a normalized Bark scale. (a) Input speech spectrum (dotted line) and the LP spectrum, (b) critical band filter-banks, (c) critical band filtered output spectrum, (d) the equal loudness curve used for pre-emphasis, (e) the auditory PSD, (f) the PLP spectrum.

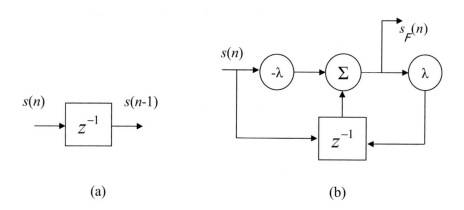

(a)   (b)

Figure 2.13: (a) The unit-delay element z^{-1} (or $W_{LP}(z)|_{\lambda=0}$) and (b) the all-pass filter, $W_{LP}(z) = (z^{-1} - \lambda)/(1 - \lambda z^{-1})$. The all-pass filter output $s_w(n) = -\lambda s(n) + s(n-1) + \lambda s_w(n-1)$.

```
%~~~~~~~~~~~~~~~~~~~~~~~~~~~~~~~~~~~~~~~~~~~~~~~~~~~~~~~~~~~~~~~~~~~~~~~~~~~~~~~~~~~~~
% Sampling-rate dependent warping coefficient.
%~~~~~~~~~~~~~~~~~~~~~~~~~~~~~~~~~~~~~~~~~~~~~~~~~~~~~~~~~~~~~~~~~~~~~~~~~~~~~~~~~~~~~
function wlp_phaseResp()

 pfs = [ 1000, 2000, 3000, 4000, 6000, 8000, 10000, 14000, 16000, ...
         20000, 30000, 32000, 44100, 50000, 70000, 100000];
 for ipfs = 1 : length(pfs)
   lamda(ipfs) = 1.0674*sqrt((2/pi)*atan(0.06583 * pfs(ipfs)/1000)) - 0.1916;
 end
 figure(1), semilogx(pfs, lamda);

 %%%
 % warped LP phase response
 %%%
 lambda = [0.9, 0.6, 0.4, 0.2, 0, -0.2, -0.4, -0.6, -0.9];
 for ic = 1 : 1 : length(lambda)
  bcoeff = [-lambda(ic), 1];
  acoeff = [1, -lambda(ic)];
  H = freqz(bcoeff, acoeff);
  figure(2), plot(unwrap(angle(H))), hold on,
 end
```

Figure 2.14: Warped LP (WLP) phase response for different values of the warping coefficient. Sampling-rate dependent warping coefficient.

Figure 2.15 (a) plots the phase response of the all-pass filter for various warping coefficients ranging from -0.9 to 0.9. For positive values of λ, the frequency response is warped toward low frequencies. On the other hand, for $\lambda < 0$, the frequency scale is unwarped. The warping coefficient depends on the sampling rate and is typically computed such that the frequency scaling mimics the auditory filter-bank characteristics. An analytical expression for λ is given by [69],

$$\lambda = 1.0674 \sqrt{\left[\frac{2}{\pi} \arctan \left(\frac{0.06583 f_s}{1000} \right) \right]} - 0.1916 . \tag{2.19}$$

Figure 2.15 (b) depicts how the warping coefficient, λ, varies with respect to the sampling rate, f_s. For narrowband speech, $\lambda = 0.4013$ and for wideband speech, $\lambda = 0.5755$.

The mathematical formulation of the conventional LP analysis (Section 2.1) can be easily extended to warped LP by replacing the unit-delay elements, z^{-1}, in Eq. (2.1), using the all-pass filter, $W_{LP}(z)$. This is given by,

$$A_W(z) = 1 - \sum_{i=1}^{L} a_{w,i} \, (W_{LP}(z))^i \tag{2.20}$$

where $A_W(z)$ is the WLP analyis filter and $a_{w,i}$ are the warped LP coefficients. The residual error, $e_w(n)$, that results from the prediction process is given by,

$$e_w(n) = s(n) - \hat{s}_w(n)$$
$$= s(n) - \sum_{i=1}^{L} a_{w,i} s_{w,i}(n) \tag{2.21}$$

where $s_{w,i}(n)$ is the inverse z-transform of $(W_{LP}(z))^i S(z)$, i.e.,

$$s_{w,i}(n) = \underbrace{w_{LP}(n) * w_{LP}(n) * \cdots * w_{LP}(n)}_{i-\text{fold}} * s(n) \tag{2.22}$$

where $w_{LP}(n)$ is the impulse response of $W_{LP}(z)$ and "*" is the convolution operator. The warped LP coefficients, $a_{w,i}$, are estimated using the least squares minimization procedure similar to the one described in Section 2.1, Eqs. (2.5) and (2.6). The resulting normal equations in the warped LP case is given by,

$$\sum_{k=1}^{L} a_{w,k} \sum_{n} s_{w,k}(n) s_{w,i}(n) = \sum_{n} s(n) s_{w,i}(n), \quad 1 \le i \le L . \tag{2.23}$$

Frequency scale warping results in an improved spectral resolution only up to a certain frequency, called turning point frequency, f_t.

$$f_t = \frac{f_s}{2\pi} \arccos(\lambda) . \tag{2.24}$$

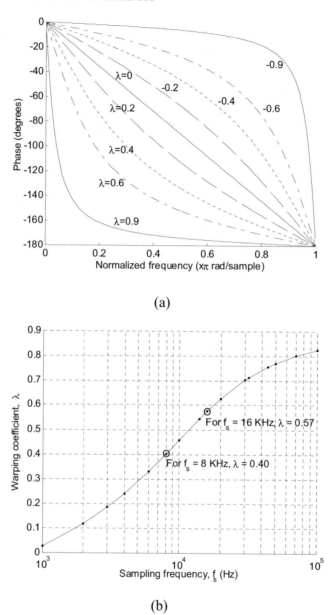

(a)

(b)

Figure 2.15: The phase response of the all-pass filter, $W_{LP}(z)$, for different values of the warping coefficient, λ. (b) Sampling rate-dependent warping coefficient. For the best match with the Bark scale, for f_s = 8 kHz, λ = 0.4013 and for f_s = 16 kHz, λ = 0.5755. Note that these λ values do not necessarily have to result in the best MOS improvements in subjective evaluations.

```
%~~~~~~~~~~~~~~~~~~~~~~~~~~~~~~~~~~~~~~~~~~~~~~~~~~~~~~~~~~~~~~~~~~~~~~~~~~~~~~~~~~~~~
%% WLP analysis
%~~~~~~~~~~~~~~~~~~~~~~~~~~~~~~~~~~~~~~~~~~~~~~~~~~~~~~~~~~~~~~~~~~~~~~~~~~~~~~~~~~~~~
function wlpCoef = wlpAnalysis(speech, fs, lpOrder)

  % Warping coefficient
  lambda = 1.0674 * (((2/pi) * atan(0.06583*fs/1000))^0.5) - 0.1916;

  % all-pass filter b and a coefficients
  b = [-lambda 1]';
  a = [1 -lambda]';

  sig = speech(:);   % speech as column vector
  sigt = sig';

  % warped auto-correlation coefficient at zero lag.
  wac(1,1) = sigt * sig;

  % Initialize the warped signal
  wrpd = sig(:);

  % Calculate warped auto-correlation coefficients.
  for i = 2 : lpOrder+1
    % warped speech
    wrpd = filter(b, a, wrpd);
    % warped autocorr
    wac(1,i) = sigt * wrpd;
  end

  % Levinson-Durbin recursion to estimate the warped LP coefficients.
  [wlpCoef, e] = levinson(wac, lpOrder);
```

Figure 2.16: Warped LP analysis.

Beyond the turning point frequency, the frequency resolution decreases. Table 2.2 presents some of the typical sampling rates and the corresponding λ and f_t values used in WLP analysis. An interesting insight from Table 2.2 is that the turning point frequency in case of narrowband speech is around 1.47 kHz. This is also evident from Figure 2.17 where the WLP fits the FFT spectrum well near lower frequencies. The spectral fitting of the WLP deteriorates steadily for frequencies greater than f_t.

   A simulation example that further illustrates the advantages of employing perceptually-driven non-uniform frequency resolution in the PLP and WLP analysis is given next. Two different input signals are considered in this example, i.e., a 20 ms wideband speech segment and a tonal signal.

   1. The FFT spectrum of a 20 ms wideband voiced speech segment is shown in Figure 2.18 (a). In this figure, we also present the tenth-order LP, WLP, and PLP spectral envelopes. Also shown are the pole frequency locations identified by these methods. It is evident that the LP places equal emphasis at all the frequencies while reducing the prediction error. On the other hand, the PLP and the WLP employs a finer resolution at low frequencies. The first four

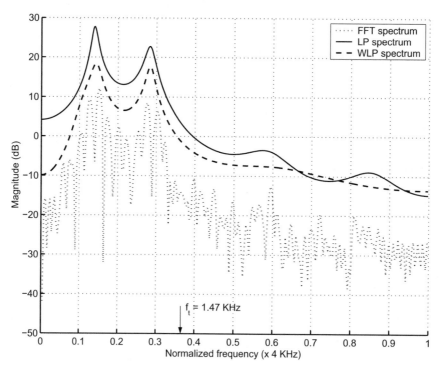

Figure 2.17: A tenth-order LP spectrum, the FFT spectrum, and the WLP spectrum of the 20 ms voiced speech segment shown in Figure 2.5 (a). For comparison purposes, the WLP spectrum shown here is unwarped along the frequency axis using $\lambda = 0.4013$. For clarity, the FFT spectrum is moved down by 10 dB.

spectral peaks are modeled reasonably well by the PLP and the WLP. We chose a tenth-order predictor for comparison purposes only.

2. Consider the signal, $s(n) = \sin(0.04\pi n) + \sin(0.1\pi n) + \sin(0.14\pi n) + \sin(0.2\pi n) + \sin(0.4\pi n) + \sin(0.6\pi n) + \sin(0.7\pi n) + \text{AWGN}$. In Figure 2.18 (b), we demonstrate how 14-th order (seven conjugate pole pairs) LP, PLP, and WLP methods model $s(n)$. The LP models the spectral peaks 2, 4, 5, 6, and 7 (counted from left to right) by placing a pole at that particular frequency. The LP fails to model well the sinusoids at frequencies 0.32 kHz and 1.12 kHz. Both the PLP and WLP, on the other hand, model the first five spectral peaks well. The tones at 4.8 kHz and 5.6 kHz, however, are not modeled well by the PLP and the WLP. This is because the frequency scale resolution of these methods decreases after the turning point frequency, $f_t=2.44$ kHz (Table 2.2). In fact, the two high frequency tones at

Table 2.2: A list of design parameters (the warping coefficient, λ, and the turning point frequency, f_t) in WLP analysis for various sampling rates, f_s.

f_s (kHz)	λ	f_t (kHz)
8	0.4013	1.47
16	0.5755	2.44
44.1	0.7654	5.0
48	0.7660	5.33

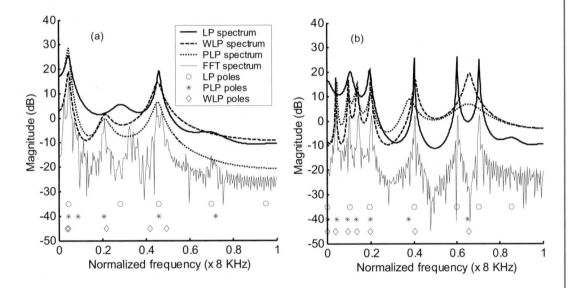

Figure 2.18: Spectral modeling of the LP, the PLP, and the WLP methods when the input signal is, (a) a wideband speech segment, (b) a sum of sinusoids signal. For clarity, a common legend is used for both the plots and is shown on the left side.

4.8 kHz and 5.6 kHz are modeled with only one pole that is placed around 5.2 kHz (see Figure 2.18 (b)).

The WLP has shown promise for reduced complexity in wideband audio coding by minimizing the need for an explicit perceptual model (e.g., [46] and references therein). The use of all-pass sections in the WLP analysis would result in delay-free loops, i.e., the prediction of the current sample is based on the linear combination of past L samples and the present sample. Computationally expensive recursive filters may be required for the WLP synthesis [77]. Details on the design of frequency-warped recursive filters can be found in [76] and [78]. The WLP sacrifices the whitening property in order to provide better modeling at some frequencies and to accommodate for the perceptual shaping of quantization noise. A drop of 1-15 dB in the spectral flatness measure associated with the WLP was reported in [46].

2.4   DISCRETE ALL-POLE MODELING

The LP error criterion in the frequency domain is given by,

$$
\varepsilon = \frac{1}{N} \sum_{m=0}^{N-1} \frac{|S(\Omega_m)|^2}{|H(\Omega_m)|^2} , \tag{2.25}
$$

where $|S(\Omega_m)|^2$ is the power spectral density (PSD) of the input speech segment, N is the number of discrete frequencies $\Omega_m = e^{-j2\pi m/N}$, and $|H(\Omega_m)|^2$ is the PSD associated with the LP synthesis filter, $1/A(z)$. The LP error criterion, Eq. (2.25), matches the autocorrelation of the continuous all-pole model to the autocorrelation of the given signal, however, without compensating for the distortion created due to spectral sampling. This yields LP spectral estimates that are biased towards the pitch harmonics [11, 49]. An example simulation that explains the sensitivity of LP estimates to high-pitched sounds is presented below. We consider the same input speech segment shown in Figure 2.5. The autocorrelation ($l = 64$ lags) of the input speech, R_{orig}, and the corresponding tenth order LP spectral model, R_{LP}, are shown in Figure 2.19 (a) and (b), respectively. Two important observations can be made from this figure,

1. the R_{LP} will always equal an aliased version of R_{orig}, and

2. R_{LP} matches closely only the first 11 (i.e., $L+1$) coefficients of R_{orig}.

To this end, Makhoul and El-Jaroudi [42] proposed a discrete all-pole (DAP) modeling that fits the LP spectral envelope to a finite set of spectral points by employing a modified LP error criterion.

The error minimization criterion employed in the DAP modeling is based on the discrete form of the Itakura-Saito (IS) distance measure [55, 79]. The IS distance measure is given by,

$$
\varepsilon_{IS} = \frac{1}{N} \sum_{m=0}^{N-1} \left[\frac{|S(\Omega_m)|^2}{|H(\Omega_m)|^2} - \ln \left(\frac{|S(\Omega_m)|^2}{|H(\Omega_m)|^2} \right) - 1 \right] . \tag{2.26}
$$

The two error criteria in Equations (2.25) and (2.26) are shown in Figure 2.20. From this figure, Eq. (2.25) gives equal weight to model errors below and above the measured spectrum. On the other hand, Eq. (2.26) gives relatively more weight to errors that occur when the model spectrum lies below the measured samples. Because of this, the estimated DAP spectrum from ε_{IS} sits on top of the FFT spectrum (see Figure 2.22).

Minimization of discrete-form of the IS measure, i.e., $\partial \varepsilon_{IS}/\partial a_i = 0$, $1 \le i \le L$, would yield a set of non-linear equations, i.e.,

$$
2 \sum_{k=0}^{L} a_k \left[r_{ss}(i - k) - \hat{r}_{lp}(i - k) \right] = 0, \ 1 \le i \le L \tag{2.27}
$$

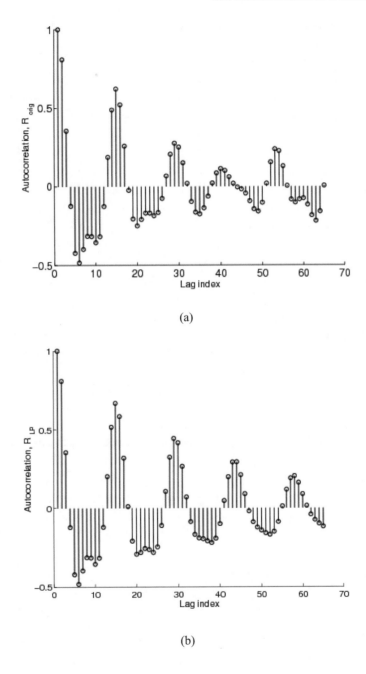

(a)

(b)

Figure 2.19: (a) Autocorrelation of the input speech for lag, l=64, and (b) autocorrelation of the 10-th order LP spectral model.

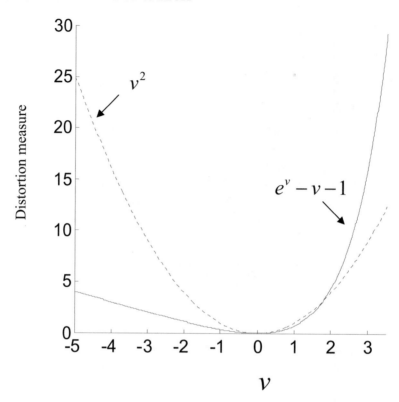

Figure 2.20: The LP and the Itakura-Saito distortion metrics, where $v = \ln\left(|S(\Omega_m)|^2 / |H(\Omega_m)|^2\right)$.

or more compactly,

$$\mathbf{R}_{ss}\,\mathbf{a} = \hat{\mathbf{R}}_{lp}\,\mathbf{a} \tag{2.28}$$

where $\hat{\mathbf{R}}_{lp}$ is the autocorrelation associated with the all-pole model sampled at the discrete frequencies, Ω_m and is given by,

$$\hat{R}_{lp}(i) = \frac{1}{N}\sum_{m=0}^{N-1}\left(\frac{1}{|A(\Omega_m)|^2}\right)\cos(\Omega_m i),\ \ 0 \leq i \leq L\ . \tag{2.29}$$

Using Eqs. (2.25) and (2.29), it can be shown that $\mathbf{R}_{ss}\,\mathbf{a} = \hat{\mathbf{h}}$, where $\hat{\mathbf{h}}$ is a column vector with time-reversed impulse response of the discrete-frequency sampled all-pole model, i.e.,

$$\hat{h}(-i) = \frac{1}{N}\sum_{m=0}^{N-1}\left(\frac{e^{-j\Omega_m i}}{A(\Omega_m)}\right),\ \ 0 \leq i \leq L\ . \tag{2.30}$$

```
%~~~~~~~~~~~~~~~~~~~~~~~~~~~~~~~~~~~~~~~~~~~~~~~~~~~~~~~~~~~~~~~~~~~~~~~~~~~~~~~~~~~~~~~~
%% Discrete all pole modeling
%~~~~~~~~~~~~~~~~~~~~~~~~~~~~~~~~~~~~~~~~~~~~~~~~~~~~~~~~~~~~~~~~~~~~~~~~~~~~~~~~~~~~~~~~
function dapLP = DAP_Analysis(s, lpOrder)

  N = length(s);  % Obtain the length of speech frame
  s = s(:);    % Make sure the speech frame is a column vector
  H2 = 10*log10(abs(fft(s, 1024)));  % for plotting purposes

  % Estimate the LP coefficients
  lpCoeff = real(lpc(s, lpOrder));

  % Convergence factor
  mu = 0.3;
  % Initialize the DAP coefficients to LP
  dapLP = lpCoeff;
  % Estimate the autocorrelation coefficients
  rr1 = xcorr(s, lpOrder);
  % Extract the second half of the autocorr coeff.
  rr = rr1(lpOrder+1 : end);
  % The Autocorrelation matrix
  RR = toeplitz(rr);
  % The inverse autocorr matrix.
  RR_inv = inv(RR);
  % Run the adaptive algorithm over 30 iterations to compute the DAP coeff.
  for iter = 1 : 1 : 30
    iter
    for i = 1: 1: lpOrder+1
      h_imp(i) = abs(mean( freqz([zeros(i,1);1], dapLP) ));
    end
    dapLP = (1-mu)*dapLP + mu * (RR_inv * (h_imp.')).';
    dapLP = dapLP / dapLP(1);
    % Plotting of DAP spectrum every iteration
    H1 = 10*log10(abs(freqz(1, dapLP)));
    figure(2), plot(H2(1:end/2+1),':'); hold on,
    plot(H1), hold off,
    pause(0.5),
  end
```

Figure 2.21: Discrete all-pole modeling Matlab program.

Note that in case of LP, $\mathbf{R}_{ss}\,\mathbf{a} = 0$, i.e., LP assumes $\hat{h}(-i) = 0,\ 0 \leq i \leq L$, which is not true for a discrete-spectrum case.

An iterative algorithm was proposed in [42] to solve for the prediction coefficients in Eq. (2.28). First, an estimate of the LP coefficients, $\mathbf{a}^{[n]}$, is obtained at iteration n. Then a new estimate, $\mathbf{a}^{[n+1]}$, is computed as follows.

$$\mathbf{a}^{[n+1]} = \mathbf{R}_{ss}^{-1}\ \hat{\mathbf{R}}_{lp}^{[n]}\mathbf{a}^{[n]}\ . \tag{2.31}$$

Rearranging the terms, we have,

$$\mathbf{a}^{[n+1]} = \mathbf{a}^{[n]} - (2\mathbf{R}_{ss})^{-1}\,\mathbf{g}^{[n]} \tag{2.32}$$

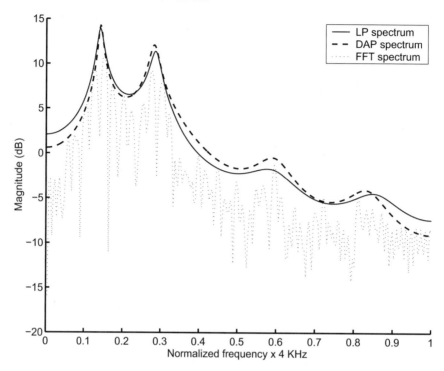

Figure 2.22: Discrete all-pole modeling. Spectral modeling performance of LP (solid line) and the DAP (dashed line). It is evident that the DAP model yields an improved spectral fitting performance. The convergence factor used in this simulation is, $\mu = 0.5$.

where $\mathbf{g}^{[n]} = 2\left[\mathbf{R}_{ss} - \hat{\mathbf{R}}_{lp}^{[n]}\right]\mathbf{a}^{[n]}$ is the gradient. From Eq. (2.32), it can be noted that $\mathbf{R}_{ss}$ is a Hessian matrix, which is a requirement for the convergence to the true solution. In order to improve the convergence speed, an update scalar, μ, is also used. The new DAP estimate can be obtained using the update equation,

$$\mathbf{a}^{[n+1]} = (1 - \mu)\,\mathbf{a}^{[n]} + \mu\mathbf{R}_{ss}^{-1}\mathbf{g}^{[n]}\,. \tag{2.33}$$

In Figure 2.22, we compare the spectral modeling performance of the conventional LP and the DAP algorithm. It is evident that the DAP model results in an improved spectral fitting performance for strongly voiced segments.

Several variants of DAP modeling have also been proposed using perceptually-weighted IS measure [42], pole-zero modeling [80], symmetric DAP COSH measure [37], and weighted MSE measure in line spectral frequency domain [38].

1. Weighted DAP Model: In order to improve the spectral matching accuracy in certain frequency regions, a frequency-dependent weighting of the ε_{WIS} error measure was also proposed [42],

$$
\varepsilon_{WIS} = \frac{1}{N} \sum_{m=0}^{N-1} |W(\Omega_m)|^2 \left[\frac{|S(\Omega_m)|^2}{|H(\Omega_m)|^2} - \ln\left(\frac{|S(\Omega_m)|^2}{|H(\Omega_m)|^2}\right) - 1 \right] \tag{2.34}
$$

where $|W(\Omega_m)|^2$ is a weighting function that typically corresponds to a perceptually-relevantscale, e.g., a Mel scale [68]. The DAP algorithm was later extended to pole-zero modeling of discrete spectra [80].

2. DAP COSH Model: Discrete spectral modeling based on the symmetric COSH distance measure, ε_{COSH} has also been proposed [37].

$$
\varepsilon_{COSH} = \frac{1}{N} \sum_{m=0}^{N-1} \cosh\left[\ln\left(\frac{|S(\Omega_m)|^2}{|H(\Omega_m)|^2}\right) \right] - 1
$$

$$
= \frac{1}{2N} \sum_{m=0}^{N-1} \left[\frac{|S(\Omega_m)|^2}{|H(\Omega_m)|^2} + \frac{|H(\Omega_m)|^2}{|S(\Omega_m)|^2} - 2 \right]. \tag{2.35}
$$

It was shown [81] that the COSH distance measure is the most preferred error criterion for discrete spectral all-pole modeling. Because the COSH distance measure is symmetric in nature, the DAP-COSH model yields improved spectral fitting both at the peaks and the valleys.

3. Weighted MSE All-Pole Model: Petrinovic [38] proposed a discrete weighted mean square all-pole modeling approach. This algorithm differs from the DAP model in two ways. First, instead of using the IS distortion measure, a weighted MSE spectral distance measure is used. Second, instead of modifying the predictor coefficients in the LPC domain as in Eqs. (2.31)–(2.33), the estimation is performed in the LSF domain. A major advantage of this method is that using the interlacing property of the roots of the symmetric and asymmetric LSPs, one can easily perform stability checks on the LP synthesis filter.

2.5   LP WITH MODIFIED FILTER STRUCTURES

In LP with modified filter structures method [63], the prediction coefficients are computed from a generalized linear representation of signal history as seen through a filter-bank. In particular, the delay elements, z^{-i}, in the conventional LP are replaced by a linear filter, $D_i(z)$ (see Figure 2.23).

$$
A_M(z) = 1 - \sum_{i=1}^{L} a_i D_i(z) . \tag{2.36}
$$

In general, the filters, $D_i(z)$, are chosen such that they represent certain specific properties of speech.

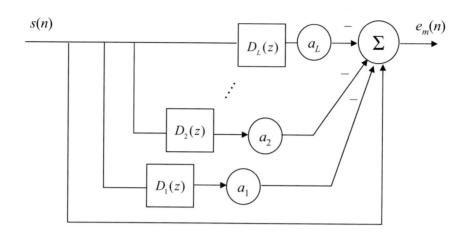

Figure 2.23: Linear prediction with modified filter structures. In this figure, $D_i(z)$ corresponds to the i-th all-pass section, $s(n)$ is the input speech sample, a_i, i=1,2,..., L are the prediction coefficients, L is the prediction order, and $e_m(n)$ is the prediction residual.

For example, $D_i(z)$ could be a long-term predictor or a filter that models the slowly time-varying spectral characteristics of the upper vocal tract. Such systems are difficult to implement because of the stability issues associated with the LP synthesis filters. By constraining to a set of linear filters that satisfy the principle of "uniform sampling," one can overcome the difficulties of controlling the stability of the synthesis filter. The principle of uniform sampling means

$$D_i(z) = [D_1(z)]^i, \ i = 1, 2, \ldots L.$$

For $D_i(z) = \left[\left(z^{-1} - \lambda\right)/\left(1 - \lambda z^{-1}\right)\right]^i$, the modified LP will result in warped LP. Other LP systems that are based on orthogonal polynomial bases (e.g., the Kautz and Laguerre models) were also reported in [63]. For example, the Kautz filter [82] is given by,

$$D_i(z) = \frac{\sqrt{1 - |\lambda_i|^2}}{1 - \lambda_i z^{-1}} \prod_{j=1}^{i} \frac{z^{-1} - \lambda_j}{1 - \lambda_j z^{-1}}. \tag{2.37}$$

The tapped-delay line model is a special case of Kautz filter for $\lambda_i = \lambda_j = 0$, $\forall i, j$. The Laguerre filter can be obtained for $\lambda_i = \lambda_j$, $\forall i, j$. A simplified form of the Laguerre filter can be implemented as a tapped all-pass line preceded by the section, $\sqrt{1 - |\lambda|^2}\big/1 - \lambda z^{-1}$. The effect of λ in Laguerre filters is same as the WLP, i.e., improved modeling toward the low frequencies for $\lambda > 0$ or more emphasis towards modeling the high frequencies for $\lambda < 0$. From Section 2.3, it can be noted that constructing synthesis filters for such WLP structures would lead to delay-free loops. Advantages of LP with modified filter structures include the possibility to enhance the spectral modeling performance at select frequency regions and the ability to incorporate *a priori* signal information into analysis-synthesis.

2.6   IIR-BASED PURE LP

In IIR-based pure LP [64], the main idea is to predict the current sample based on the IIR-filtered past samples (Figure 2.24). For a specific class of filter structures, e.g., the Kautz [82] and Laguerre

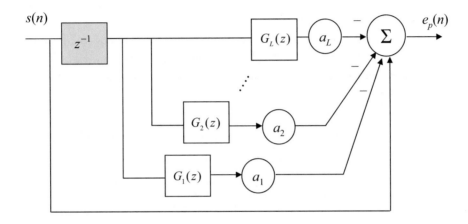

Figure 2.24: IIR-based pure LP. Because the predictor estimates the current sample based on the past samples only, this case is referred to as 'pure' linear prediction. In this figure, $G_i(z)$ are stable, causal IIR filters. The use of the unit delay element z^{-1}, at the front end, enables this structure to avoid the problems associated with the delay-free loops.

filters [83], and with appropriate input signal windowing, the IIR-based pure LP yields stable synthesis filters [64]. The main idea is to replace $D_i(z)$ in Eq. (2.36) with causal, stable IIR filters. This is given by,

$$D_i(z) = z^{-1}G_i(z) \tag{2.38}$$

where $G_i(z)$ are stable and causal IIR filters.

2.7   LP MODELING BASED ON WEIGHTED SUM OF LSP

An all-pole modeling approach that is based on the weighted-sum of LSP (WLSP) polynomials is proposed in [65], see Figure 2.25. In particular, define the prediction polynomial,

$$A(z, \alpha) = \alpha P(z) + (1 - \alpha) Q(z) \tag{2.39}$$

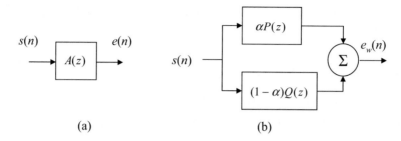

(a)   (b)

Figure 2.25: Different interpretations of the LP analysis modeling, (a) conventional LP, (b) weighted sum of LSP polynomials. In this figure, $P(z)$ and $Q(z)$ are the symmetric and asymmetric LSP polynomials. A suitable value of the weighting parameter, α, improves the autocorrelation matching between the all-pole model and the input speech.

where $P(z)$ and $Q(z)$ are the symmetric and asymmetric polynomials, and α is the weighting parameter. The symmetric and asymmetric LSP polynomials are computed as follows,

$$P(z) = \frac{A(z) + z^{-(L+1)}A(z^{-1})}{1 + z^{-1}} \qquad (2.40)$$

$$Q(z) = \frac{A(z) - z^{-(L+1)}A(z^{-1})}{1 - z^{-1}} . \qquad (2.41)$$

The roots of $P(z)$ and $Q(z)$ lie on the unit circle and alternate, provided the poles of the LP synthesis filter, $1/A(z)$, are inside the unit circle. Figure 2.26 shows a z domain plot depicting the interlacing property of the roots of the symmetric and asymmetric LSP polynomials.

An optimal value of α in Eq. (2.39) will improve the autocorrelation matching between the all-pole model and the input speech. This, in turn, would yield all-pole spectra with an increased dynamic range between formant peaks and spectral valleys. Unfortunately, there is no closed-form solution to estimate an optimal value for the weighting parameter, α. An iterative brute-force approach was suggested [65]. Note that for $\alpha = 0.5$, the WLSP will yield the conventional all-pole model.

Let us consider the input speech segment shown in Figure 2.5 (a) and examine how the LP spectral modeling varies for different values of α. Figure 2.28 (a) shows the WLSP modeling for $\alpha = 0.3, 0.5$, and 0.7. In order to better understand the underlying concepts of WLSP, let us consider the root locus plot shown in Figure 2.28 (b). In this figure, the dashed line indicates the root track of a second-order LP $A(z, \alpha)$ [Eq. (2.39)] as a function of α. Note that for $\alpha = 0$, the root of $A(z, \alpha)$ corresponds to that of an asymmetric polynomial, $Q(z)$ (shown as $\alpha = 0$, $Q(z)$ root). Similarly, for $\alpha = 1$, the root of $A(z, \alpha)$ corresponds to that of a symmetric polynomial, $P(z)$ (shown as $\alpha = 1$, $P(z)$ root). As α is increased from 0 to 1, the root of $A(z, \alpha)$ moves along the root locus, and for a certain value, it yields the best autocorrelation matching between the all-pole model and the

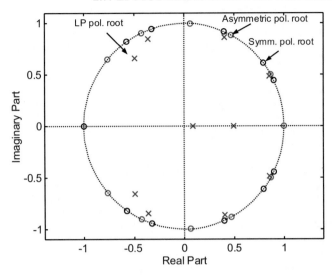

Figure 2.26: A z domain plot depicting the interlacing property of the roots of the symmetric and asymmetric LSP polynomials. The pole locations of the LP synthesis filter, $1/A(z)$, are shown as crosses 'x,' the roots of the symmetric and asymmetric LSP polynomials are shown as black and red circles, respectively. Note the interlacing nature of black and red circles and they always lie on the unit circle. Also, note that if a pole is close to the unit circle, the corresponding LSPs will be very close to each other.

```
%~~~~~~~~~~~~~~~~~~~~~~~~~~~~~~~~~~~~~~~~~~~~~~~~~~~~~~~~~~~~~~~~~~~~~~~~~~~~~~~~~~~~~~~
%% LP Modeling based on weighted sum of LSP
%~~~~~~~~~~~~~~~~~~~~~~~~~~~~~~~~~~~~~~~~~~~~~~~~~~~~~~~~~~~~~~~~~~~~~~~~~~~~~~~~~~~~~~~
function weightedLP = WghtSUM_LSP(s, lpOrder)

  N = length(s);  % Obtain the length of speech frame
  s = s(:);
  H2 = 10*log10(abs(fft(s, 1024)));  % for plotting purposes

  % LP analysis
  lpCoeff = real(lpc(s, lpOrder));

  % Symmetric LSP polynomial
  pp = [lpCoeff, 0] + [0, fliplr(lpCoeff)];

  % Asymmetric LSP polynomial
  qq = [lpCoeff, 0] - [0, fliplr(lpCoeff)];

  for alpha = [0.3, 0.4, 0.7]
    weightedLP = alpha*pp + (1-alpha)*qq;
    % Plotting of DAP spectrum every iteration
    H1 = 10*log10(abs(freqz(1, weightedLP)));
    figure(2), plot(H2(1:end/2+1),':'); hold on,
    plot(H1), hold off,
    pause,
  end
```

Figure 2.27: LP using weighted-sum of LSP polynomials.

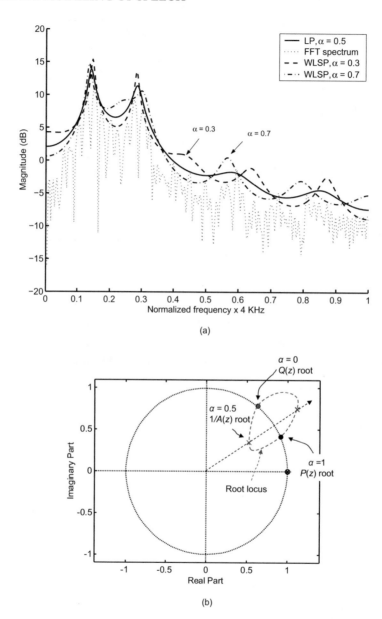

Figure 2.28: (a) All-pole modeling based on weighted sum of LSP polynomials when the weighting factor, $\alpha = 0.3, 0.5$, and 0.7. (b) The root locus plot of $A(z, \alpha)$. The red dashed line corresponds to the root track of $A(z, \alpha)$ as a function of α. Note that for $\alpha = 0$, the root corresponds to $Q(z)$ (shown as red circle). Similarly, for $\alpha = 1$, the root corresponds to $P(z)$ (shown as black circle).

input speech. In some sense, the autocorrelation matching problem that we are discussing here is analogous to the one that we addressed in Section 2.4. In order to satisfy the constraint that the prediction residual energy should decrease for increasing model order, the value of α is restricted between 0.2929 and 1. When $\alpha = 0.3$, $A(z) = 0.3P(z) + 0.7Q(z)$, which means that more weight is given to the asymmetric polynomial in estimating the LP. The root of $1/A(z)$ will move closer to that of the $Q(z)$ along the root locus (red dashed line in Figure 2.28).

2.8   LP WITH LOW-FREQUENCY EMPHASIS

In order to provide better spectral modeling at the low frequency formants, an all-pole modeling approach, based upon the symmetric linear prediction [84], was proposed in [39]. The algorithm (Figure 2.29) is called linear prediction with low-frequency emphasis (LPLE). The LPLE approach

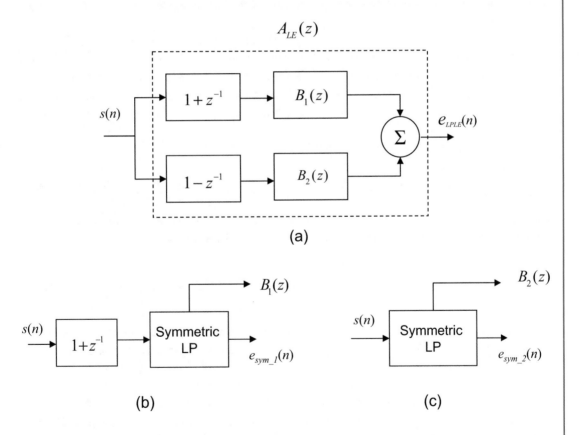

Figure 2.29: (a) Symmetric LP modeling with low-frequency emphasis. (b) Steps to compute $B_1(z)$, (c) Steps to compute $B_2(z)$. In this figure, $B_1(z)$ and $B_2(z)$ are symmetric linear predictors of order L computed from the input speech filtered through pre-filters, $(1+z^{-1})$ and 1, respectively.

works on the fact that the LSP symmetric and asymmetric polynomials are two optimal predictors [85] that minimize the prediction error energy subject to the constraint that the zeros of the predictor are restricted to the unit circle [84]. For an even-order LP, the prediction polynomial, $A(z)$, is computed as,

$$A_{LE}(z) = 0.5\,[P(z) + Q(z)]$$
$$= 0.5\left[(1 + z^{-1})B_1(z) + (1 - z^{-1})B_2(z)\right] \tag{2.42}$$

where $B_1(z)$ and $B_2(z)$ are transfer functions of the symmetric linear predictors of order L computed from the input speech filtered through pre-filters, $(1+z^{-1})$ and 1, respectively. Figure 2.29 (b, c) further elaborate the steps employed to compute $B_1(z)$ and $B_2(z)$. Note also that in Figure 2.29 (c), if $B_2(z)$ was computed from the input speech filtered through a pre-filter $(1-z^{-1})$, the resulting $A_{LE}(z)$ is equivalent to $A(z)$.

Figure 2.30 (a) shows the spectral fitting of the conventional LP (solid line) and the LPLE algorithm (dashed and dashed-dot lines for prediction orders, L = 10 and 8). Figure 2.30 (b) shows the magnitude spectrum of the two symmetric linear predictors, $B_1(z)$ and $B_2(z)$ associated with the LPLE model. The LPLE algorithm with model order, L=10, overemphasizes the low frequencies Figure 2.30. It is evident that for L=8, the LPLE algorithm estimates the first two formants accurately.

2.9 LP IN CASCADE-FORM

Motivated by the need to directly estimate the roots of the prediction filter, especially, in speech analysis applications, LP has been studied in cascade form [66], [86, 87]. Adaptive LP methods that are based on the FIR cascade structure have received attention in lossless compression of audio [88]–[90]. The LP analysis filter in cascade form (see Figure 2.31) is given by,

$$A(z) = \prod_{i=1}^{p}\left(1 - 2r_i \cos\theta_i z^{-1} + r_i^2 z^{-2}\right) \tag{2.43}$$

where (r_i, θ_i) denote the i-th pole location in polar coordinates and p is the total number of poles. The parameters, r_i and θ_i, are related directly to the source-system physical attributes of interest. For example, the center frequency, $f_i = \theta_i f_s / 2\pi$, and bandwidth, $BW_i = f_s \ln(r_i)/\pi$, correspond to the formant frequency and the resonance strength, where f_s is the sampling frequency.

Figure 2.31 shows a general structure for the LP in cascade form. The conventional direct-form LP was reformulated to directly compute the roots of the predictor polynomial in an iterative manner. In particular, the pole angles, θ_i, and the corresponding pole amplitudes, r_i, were estimated iteratively by solving p simultaneous non-linear equations Eqs. (2.45) and (2.46). One important point is that, even though non-linear, these equations possess a unique solution if the corresponding normal equations for the direct-form have a unique solution, as they result from minimization of

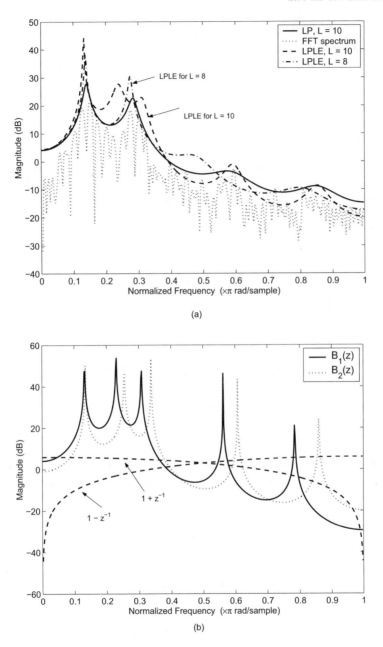

Figure 2.30: (a) LP with low-frequency emphasis (LPLE) spectral modeling. (b) Magnitude spectra of $B_1(z)$ and $B_2(z)$. The LPLE algorithm is well-suited for spectral fitting where (1) the prediction order is small and (2) more emphasis needs to be placed on modeling the low frequencies.

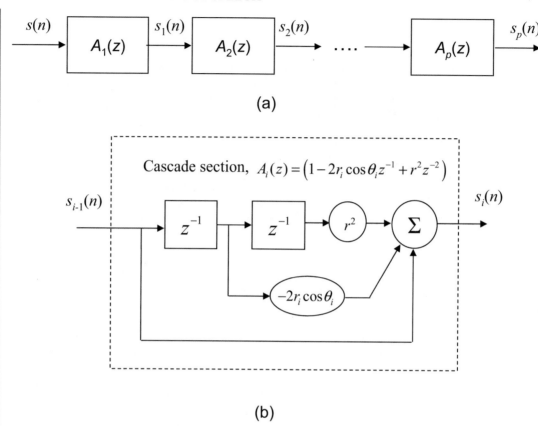

(a)

(b)

Figure 2.31: (a) LP in cascade form, $A(z) = \prod_{i=1}^{p} A_i(z)$, (b) the i-th cascade section $A_i(z) = \left(1 - 2r_i \cos \theta_i z^{-1} + r_i^2 z^{-2}\right)$. For notational convenience, we represented the prediction error $e(n)$ as $s_p(n)$.

the same MSE norm [66]. The prediction error, $E(z)$, can be defined as,

$$E(z) = S(z) \prod_{i=1}^{p} \left(1 - 2r_i \cos \theta_i z^{-1} + r_i^2 z^{-2}\right). \tag{2.44}$$

The gradient components, $\partial e(n)/\partial \theta_i$ and $\partial e(n)/\partial r_i$, are computed as follows:

$$\frac{\partial e(n)}{\partial \theta_i} = g_{1,i}(n) 2r_i \sin(\theta_i) \tag{2.45}$$

$$\frac{\partial e(n)}{\partial r_i} = \frac{1}{r_i} \left[-2r_i \cos(\theta_i) g_{1,i}(n) + 2r_i^2 g_{2,i}(n)\right] \tag{2.46}$$

where $g_{1,i}(n)$ are obtained by passing the input sequence, $s(n)$, through all but the i-th cascade filter section, $A_i(z) = \left(1 - 2r_i \cos\theta_i z^{-1} + r_i^2 z^{-2}\right)$, and then delaying the output by l samples. Note that the r and θ directions are orthogonal, which allows to adjust these parameters independently, and the cross-correlation terms within each section are typically ignored. Using the aforementioned gradient components, the update equations for the pole angle and the pole amplitude are given by,

$$\Delta\theta_i = \frac{-\mu_1}{2r_i \sin(\theta_i)} \frac{\sum_n e(n)g_{1,i}(n)}{\sum_n g_{1,i}^2(n)} \tag{2.47}$$

$$\Delta r_i = -\mu_2 r_i \frac{\sum_n e(n)\left[-2r_i \cos(\theta_i)g_{1,i}(n) + 2r_i^2 g_{2,i}(n)\right]}{\sum_n \left[-2r_i \cos(\theta_i)g_{1,i}(n) + 2r_i^2 g_{2,i}(n)\right]^2} \tag{2.48}$$

where μ_1 and μ_2 are the convergence factors.

Figure 2.32 shows an example Matlab implementation where the pole amplitudes are estimated using cascade form LP. The update equation for, r_i, given in (2.48), can be compactly

```
%~~~~~~~~~~~~~~~~~~~~~~~~~~~~~~~~~~~~~~~~~~~~~~~~~~~~~~~~~~~~~~~~~~~~~~~~~~~~~~~~~~~~~~
%% Cascade-form LP algorithm for pole amplitude estimation
%~~~~~~~~~~~~~~~~~~~~~~~~~~~~~~~~~~~~~~~~~~~~~~~~~~~~~~~~~~~~~~~~~~~~~~~~~~~~~~~~~~~~~~
function [cas_lpCoeff] = cascadeLP(speech, prev_frame, lpOrder)

  % speech : speech samples from current frame [N x 1]
  % prev_frame: includes samples from previous frame for covariance matrix calculation.
  speech = speech(:);
  s = [prev_frame(end-lpOrder-1:end), speech];
  N = length(speech);  % Obtain the length of speech frame

  % Estimate the covariance matrix
  for ip = 1 : 1 : lpOrder+1
    for jp = 1 : 1 : lpOrder+1
      phi(ip,jp)=sum(s(lpOrder+1-ip+1:lpOrder+1-ip+N).*s(lpOrder+1-jp+1:lpOrder+1-jp+N));
    end
  end


  lpCoeff = lpc(speech, lpOrder);  % LP coefficients
  lpRoots = roots(lpCoeff)';  % Roots of LP polynomial
  W = sort(angle(lpRoots));   % pole frequenices (theta).
  W = W((lpOrder/2)+1 : lpOrder);

  Hlp= freqz(speech, 1);    % Freq. Spectrum of speech
  Hlp_ener = sum(abs(Hlp))/512;
  Hhat_ener = [];
  W_i = W.';      % Pole frequencies obtained from LP polynomial
  R_i = 0.5*ones(lpOrder/2,1); % Initialize the Cascade LP root amplitudes
  mu = 0.2;  % Convergence factor
```

Figure 2.32: Cascade form LP pole amplitude estimation. *Continues.*

```matlab
for iter = 1 : 1 : 70
  fprintf('%d ', iter)

  % Impulse response of H(z)
  [xc, yc] = pol2cart(W_i', R_i');  % Convert form Polar to Cartesian form
  ppq = [ [xc + j*yc], [xc - j*yc] ];
  bcc = poly(ppq);    % Construct the polynomial from roots
  h = filter(bcc, 1, [1, zeros(1,lpOrder)]);  % Impulse response of cascade form LP filter

  for i_c = 1 : 1 : lpOrder/2
    wi = W_i;       ri = R_i;
    wi(i_c) = 0;    ri(i_c) = 0;

    % Construct the polynomial from all roots except [ri(ic), wi(ic)]
    [xcd, ycd] = pol2cart(wi', ri');
    ppqd = [ [xcd + j*ycd], [xcd - j*ycd] ];
    bccd = poly(ppqd);

    % Impulse responses of H1(z) and H2(z)
    h1 = filter([0, bccd], 1, [1, zeros(1,lpOrder)]);
    h2 = filter([0, 0, bccd], 1, [1, zeros(1,lpOrder)]);

    % The gradient parameter
    xeta = -2*R_i(i_c)*cos(W_i(i_c))*h1 + 2 * R_i(i_c) * R_i(i_c) * h2;

    % Numerator and Denominator of Eq. (2.49)
    Numer = (h * phi(1 : end, 2 : end-1)) * transpose(xeta(2 : end-1)) ;
    Denom = (xeta(2:end-1)*phi(2 : end-1, 2 : end-1) ) * transpose( xeta(2 : end-1) ) ;

    r_update = -mu * R_i(i_c) * Numer / Denom;
    R_i(i_c) =  abs(R_i(i_c) + r_update);
  end
  [xc, yc] = pol2cart(W_i', R_i');
  ppq = [ [xc + j*yc], [xc - j*yc] ];
  cas_lpCoeff = poly(ppq);

  % Plotting
  % LP spectrum
  H1 = freqz(1, lpCoeff, 512);
  % cascade LP spectrum
  H2 = freqz(1, cas_lpCoeff, 512);
  figure(1),
  plot([1/512:1/512:1], 10*log10(abs(H1)),[1/512:1/512:1], 10*log10(abs(H2)), 'k');
  pause(0.1),
end
```

Figure 2.32: *Continued.* Cascade form LP pole amplitude estimation.

represented as,

$$\Delta r_i = -\mu_2 r_i \frac{\sum_{k=0}^{2p} \sum_{l=1}^{2p-1} \hat{h}(k)\zeta_i(l)\phi(k,l)}{\sum_{k,l=1}^{2p-1} \zeta_i(k)\zeta_i(l)\phi(k,l)}, \quad 1 \le i \le p \tag{2.49}$$

where μ_2 is the convergence factor (e.g., 0.2-0.4), $\phi(k,l)$ denotes the covariance matrix of order $[2p+1, 2p+1]$ associated with the input speech segment, $\hat{h}(k)$ is the impulse response of the cascade-form LP analysis filter, and the gradient-related parameter, $\zeta_i(n)$ is given by,

$$\zeta_i(n) = -2r_i \cos(\theta_i) h_{1,i}(n) + 2r_i^2 h_{2,i}(n) \tag{2.50}$$

and $h_{m,i}(n), m = 1, 2$ denote the impulse response of $H_{m,i}(z)$,

$$H_{m,i}(z) = \frac{z^{-m} H(z)}{(1 - 2r_i \cos\theta_i z^{-1} + r_i^2 z^{-2})}. \tag{2.51}$$

The update Equation (2.49) is carried out as a block-based approach. The covariance matrix, $\phi(k,l)$, is computed only once every frame and the impulse response, $\hat{h}(k)$, is updated every iteration. The stability of the cascade-form LP filter can be easily checked for $r_i < 1$.

Figure 2.33 (a) shows the all-pole spectra computed using the tenth-order conventional LP and the cascade-form LP with $\mu = 0.1, 0.2$. The normalized error energy (NEE) convergence curves obtained for $\mu = 0.1, 0.2$ are shown in Figure 2.33 (b). The NEE is estimated using,

$$NEE^{[n]} \triangleq \frac{\sum_{m=0}^{N-1} \left| H_{ref}(\Omega_m) - \hat{H}^{[n]}(\Omega_m) \right|^2}{\sum_{m=0}^{N-1} \left| H_{ref}(\Omega_m) \right|^2} \tag{2.52}$$

where $H_{ref}(\Omega_m)$ denotes the DFT of the 20-th order direct-form LP synthesis filter, $\hat{H}^{[n]}(\Omega_m)$ represents the estimated 10-th order all-pole filter using cascade-form LP at the n-th iteration, and N (e.g., 512) is the number of discrete frequencies, $\Omega_m = e^{-j2\pi m/N}$. Jackson and Wood [66] have pointed out that an appropriate choice of μ would ultimately guarantee convergence to true values. Typically, μ values in the range of 0.2-0.4 resulted in good convergence rates. The cascade-form LP was used in formant tracking applications and as an adaptive compensator for fluid-filled catheter pressure waveforms [66].

2.10   LP COEFFICIENTS QUANTIZATION

In low bit rate coding, the LP coefficients and the residual must be efficiently quantized. Because the direct-form LP coefficients do not have adequate quantization properties, transformed LP coefficients are typically quantized. First-generation vocoders such as the LPC10e [104] and the IS-54 VSELP [122] quantize reflection coefficients (RC). The reflection coefficients are given by,

$$k_m = a_m(m), \quad m = 1, 2, \ldots L \tag{2.53}$$

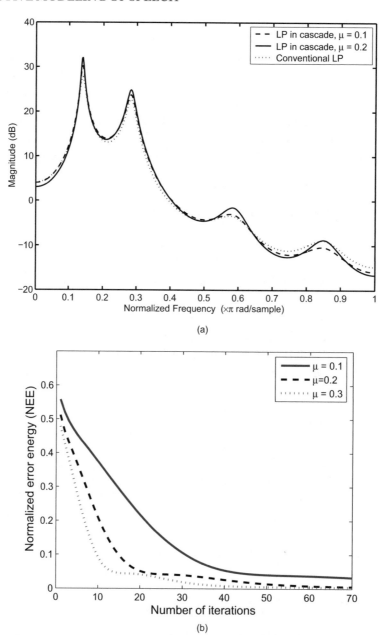

Figure 2.33: (a) Tenth-order all-pole spectra obtained using the conventional LP and the cascade-form LP with $\mu = 0.1, 0.2$. (b) Normalized error energy curves depicting the convergence characteristics of the gradient-descent algorithm, Eqs. (2.47) and (2.48).

where L is the linear predictor order. The reflection coefficients, k_m, are obtained as a by-product of the Levinson-Durbin recursion (see Matlab Program Figure 2.4). The bounding restriction, $|k_m| \leq 1$, is a necessary and sufficient condition for checking the stability of the LP synthesis filters. Hence, the reflection coefficients make an ordered set of parameters within the range of -1 to $+1$. This makes the RC to be more suitable for quantization than the direct form LP.

Furthermore, the RCs can be transformed into other equivalent set of parameters that are more robust to quantization errors such as the partial correlation (PARCOR) coefficients. The negated reflection coefficients, $-k_m$, are called PARCOR coefficients as it represents the normalized correlation between $s(n)$ and $s(n-m)$ with the correlation between $s(n-1), \ldots, s(n-m+1)$ removed.

The reflection coefficients are sensitive to quantization errors when the poles of the LP filter lie close to the unit circle. Hence, a non-linear transformation is performed to expand the region near $|k_m| = 1$, where k_m is the m-th RC. This non-linear transformation results in log area ratio (LAR) parameters.

$$\text{LAR}(m) = \log\left(\frac{1 + k_m}{1 - k_m}\right), \quad m = 1, 2, \ldots L. \tag{2.54}$$

The log area ratios and the inverse sine transformation have been used in the early GSM 6.10 algorithm [162] and in the skyphone standard [163].

Most recent LP-based cellular standards quantize line spectrum pairs (LSPs) [15] or immittance spectrum pairs (ISPs) [45, 62]. The main advantage of the LSPs is that they relate directly to frequency-domain information, and hence, they can be encoded using perceptual criteria. The line spectral frequencies, ω_m, and the line spectral pairs, q_m, are related as follows:

$$q_m = \cos(\omega_m), \quad m = 1, 2, \ldots L. \tag{2.55}$$

LSPs have a bounded range ($|q_m| \leq 1$), a sequential ordering of parameters, and a simple check for stability. For the L-th order LP filter, the LSP coefficients are defined as the roots of the sum and the difference polynomials, given by, $P(z)$ and $Q(z)$, respectively.

$$P(z) = \frac{A(z) + z^{-(L+1)}A(z^{-1})}{1 + z^{-1}} \tag{2.56}$$

$$Q(z) = \frac{A(z) - z^{-(L+1)}A(z^{-1})}{1 - z^{-1}}. \tag{2.57}$$

The roots of $P(z)$ and $Q(z)$ lie on the unit circle and alternate, provided the poles of the LP synthesis filter, $1/A(z)$, are inside the unit circle. These are the two important features of the line spectral frequencies that made them popular. The polynomial $P(z)$ is symmetric (sum polynomial), and $Q(z)$ is asymmetric (difference polynomial). Each of these polynomials has a set of $L/2$ complex conjugate pairs of zeros that lie on the unit circle. The roots, q_m, are in the cosine domain. Figure 2.26 shows the z-plane plot LP synthesis filter, $1/A(z)$, and the roots of LSP polynomials. From Figure 2.26, it can be noted that the roots of LP synthesis filter are inside the unit circle. Also, the roots of the

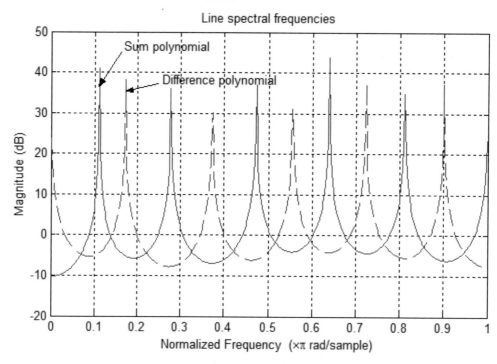

Figure 2.34: Plot depicting the alternating nature of the roots of sum and difference polynomials.

LSP polynomials lie on the unit circle and are interlaced. Figure 2.34 shows the alternating nature of the roots of sum and difference LSP polynomials.

Figure 2.36 shows the quantization distortion performance of various LP parameter transformations. In this example, a 10-th order LP is estimated from a 10 ms speech frame sampled at 16 kHz. A 4-bit quantizer is applied to the LP coefficients. Figure 2.36 (a) shows the LPC unquantized spectrum (solid line) and the 4-bit LPC quantized spectrum (dashed line). The quantization artifacts are more evident (see around 5.6 kHz in Figure 2.36 (a)) when LP coefficients are directly quantized. Figure 2.36 (b) shows the LPC spectrum from the quantized RC. Similarly, Figure 2.36 (c) and (d) show the LPC spectra obtained from the quantized LAR and the quantized LSF, respectively. The LSF and LAR coefficients show relatively better performance towards quantization artifacts than the direct-form LP coefficients. In some of the recent speech coding standards [5], the line spectrum pairs are quantized using vector quantization (VQ) methods such as the split-VQ techniques [141].

```
%~~~~~~~~~~~~~~~~~~~~~~~~~~~~~~~~~~~~~~~~~~~~~~~~~~~~~~~~~~~~~~~~~~~~~~~~~~~~~~~~~~~~~
%% LP parameter transformations example
%~~~~~~~~~~~~~~~~~~~~~~~~~~~~~~~~~~~~~~~~~~~~~~~~~~~~~~~~~~~~~~~~~~~~~~~~~~~~~~~~~~~~~
function [lp, lsf, rc] = lpParameterTrans(speech, lpOrder)

  % LP coefficients
  lpCoeff = lpc(speech, lpOrder);

  % Matlab Signal processing Toolbox has several LP parameter transformation functions.

  % LP to RC
  RC_Params = poly2rc(lpCoeff);

  % RC to LAR
  LAR_Params = rc2lar(RC_Params);

  % LP to LSF
  LSF_Params = poly2lsf(lpCoeff);
```

Figure 2.35: Various LP parameter transformations.

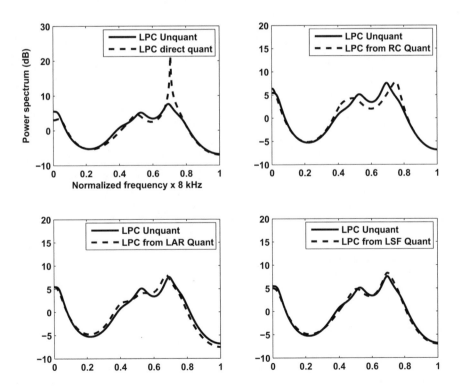

Figure 2.36: Performance of various LP parameter transformations towards quantization effects.

2.11   SUMMARY

A review of linear prediction modeling of speech is given in this Chapter. In particular, we studied the conventional LP, perceptual LP, warped LP, discrete all-pole modeling, and LP in cascade-form methods. Also some of the recent advancements such as LP with low-frequency emphasis, LP with modified filter structures, IIR-based pure LP, and LP modeling based on weighted-sum of LSP polynomials were presented. Finally, various LP parameter transformations that are common in speech coding algorithms are presented.

CHAPTER 3

Perceptual Modeling of Speech

Signal compression invariably introduces quantization noise in the original signal. However, certain kinds of quantization distortion are inaudible because of the limitations associated with the human auditory perception. Exploiting the masking properties of the human ear could yield improved coding gains. For example, a simple perceptual weighting filter that shapes the quantization noise according to the masking properties of the human ear became standard practice in speech coding [2]. In audio coding, the perceptual models are much more elaborate. Auditory models [43, 44] are now an integral part of several audio coding standards, e.g., the ISO/IEC MPEG-1 layer III (MP3) [21], the Dolby AC-3 [24], and the DTS [25, 26]. Use of elaborate auditory models in low bit rate speech coders also received interest, e.g., [135]–[137]. Advantages that can be gained by integrating auditory models in low bit rate vocoders depend on 1) the achievable coding gains, 2) computational overhead, and 3) the accuracy of these models to mimic the human performance. This Chapter focuses on how certain perceptual features from the cochlear response are processed to obtain a useful representation of the signal [131] [138, 139]. Such representations include the masked threshold [44], the ensemble interval histogram [131], the auditory excitation pattern [70, 132], the masking pattern [43], and the loudness [133].

1) Masked threshold [43] represents the masking contribution due to the existence of a tone/noise masker. Masked threshold of a tone or noise gives information on the sound-pressure-level (SPL), below which all other sounds are rendered inaudible.

2) An auditory excitation pattern (AEP) [70] can be defined as the effective energy spectrum reaching the cochlea. It describes the auditory stimulation caused by an audio signal. In terms of the filter bank analogy, the AEP can be thought of as the output of each auditory filter as a function of filter center frequency.

3) Masking pattern [43] describes how the hearing threshold of a tone or noise masker varies across the frequency. Masking patterns are useful to study the spread of masking across the adjacent critical bands.

4) The loudness [133] can be defined as the intensity of sound. Measured in Sones, the overall loudness of a given sound is proportional to the total area under the specific loudness pattern, L_s. The specific loudness is proportional to the compressed excitation pattern, i.e., $L_s \propto \left[E_p \right]^{\alpha}$, where E_p is the excitation pattern, and the compression-factor α is around 0.3.

The aforementioned auditory representations have found applications in: speech and audio coding standards [140, 141], the ITU-T perceptual evaluation of speech and audio quality (PESQ and PEAQ) standards [142, 143], residual modeling in music analysis/synthesis [144], digital audio watermarking [141], psychoacoustic matching pursuits for audio [145], speech enhancement [146], and sinusoidal component selection [103, 139].

This Chapter is organized as follows. In Section 3.1, we review the perceptual weighting techniques used in speech coding standards. In Section 3.2, we describe how a signal is processed in the human auditory system. In Section 3.3, various masking phenomena are described. A review of psychoacoustic models (e.g., [21, 147, 148]) is included in Section 3.4. Sections 3.5 and 3.6 present a step-by-step procedure to compute auditory excitation pattern and perceptual loudness, respectively.

3.1   PERCEPTUAL WEIGHTING FILTER

The prediction error from the LP analysis is modeled and encoded using a closed-loop analysis or analysis-by-synthesis (A-by-S) process. In A-by-S CELP systems, the excitation sequence that minimizes the "perceptually-weighted" mean-square-error (MSE) between the input and reconstructed speech is selected from a codebook. The perceptual weighting filter (PWF), $W(z)$, shapes the prediction error such that quantization noise is masked by the high-energy formants. The PWF is given by,

$$W(z) = \frac{A(z/\gamma_1)}{A(z/\gamma_2)} = \frac{1 - \sum_{i=1}^{L} \gamma_1^i a_i z^{-i}}{1 - \sum_{i=1}^{L} \gamma_2^i a_i z^{-i}} \tag{3.1}$$

where γ_1 and γ_2 are the adaptive weights and $0 < \gamma_2 < \gamma_1 < 1$. Typically, γ_1 ranges from 0.94 to 0.98; and γ_2 varies between 0.4 and 0.7, depending upon the tilt or the flatness characteristics associated with the LPC spectral envelope. The PWF, $W(z)$, is adaptive, and its frequency spectrum is a smoothed version of the inverse vocal tract spectrum (Figure 3.1). Hence, the role of $W(z)$ is

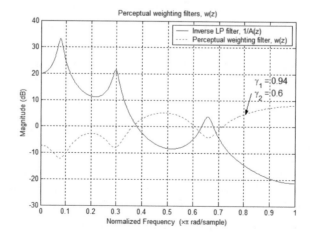

Figure 3.1: Perceptual weighting filter, $W(z)$ and LP synthesis filter, $1/A(z)$.

to de-emphasize the error energy in the formant regions [1]. This de-emphasis strategy is based on the fact that in the formant regions quantization noise is partially masked by speech. For $\gamma < 1$, the roots of the LP synthesis filter, $1/A(z/\gamma)$, move towards the origin of the unit circle as shown in Figure 3.2.

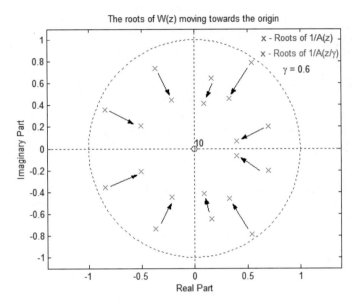

Figure 3.2: Pole locations of $1/A(z)$ and $1/A(z/\gamma)$, for $\gamma = 0.6$.

The PWF, $W(z)$, in Eq. (3.1) works well with narrowband signals and when speech spectrum does not exhibit a strong spectral tilt. However, it is not suitable for efficient perceptual weighting of wideband signals. In ITU-T G.722.2 wideband speech coding standard [45], a modified PWF that decouples the formant weighting from the spectral tilt was proposed. In particular, first a pre-emphasis filter, $\left(1 - \gamma_2 z^{-1}\right)$ is applied to the input speech, then LP coefficients are estimated from the pre-emphasized speech. While the pre-emphasis reduces the spectral tilt, the $W_{WB}(z)$ performs the perceptual weighting.

$$W_{WB}(z) = \frac{A(z/\gamma_1)}{\left(1 - \gamma_2 z^{-1}\right)} \tag{3.2}$$

where $\gamma_2 = 0.6$, and γ_1 ranges from 0.94 to 0.98. Subjective quality evaluation [115] showed that the PWF, $W_{WB}(z)$, in Eq. (3.2) that includes a pre-emphasis filter and a modified PWF is more efficient for encoding wideband signals than the PWF in (3.1).

3.2   THE HUMAN AUDITORY SYSTEM

A physiological model of the human auditory system (HAS) is shown in Figure 3.3. The human

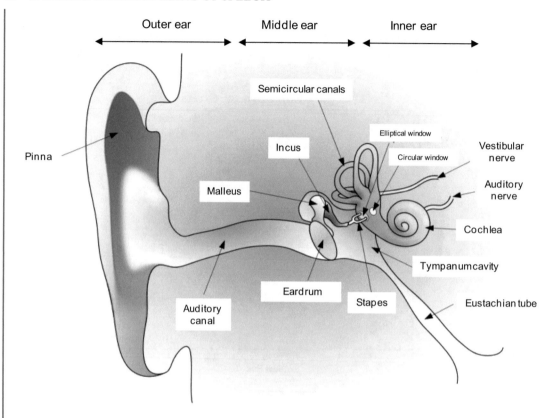

Figure 3.3: A physiological model of the human auditory system (after [149]). The outer ear consists of the pinna and the external auditory canal. The middle ear consists of the ear drum and a mechanical transducer. The inner ear consists of the cochlea, a fluid-filled chamber partitioned by the basilar membrane and the auditory nerve.

auditory system can be divided into three parts:

1) The outer ear consists of the pinna and the external auditory canal.

2) The middle ear consists of the ear drum and a mechanical transducer that comprises of the malleus, the incus, and the stapes.

3) The inner ear consists of the cochlea, a fluid-filled chamber partitioned by the basilar membrane and the auditory nerve.

The sound undergoes a nonlinear transformation through the outer and middle ear. Figure 3.4 shows the frequency responses depicting the effective attenuation associated with the outer and middle ear.

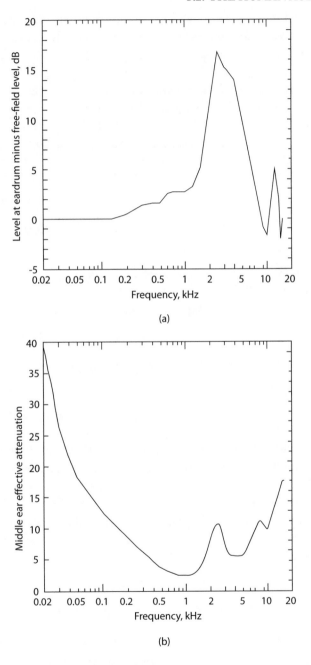

(a)

(b)

Figure 3.4: A frequency response depicting the effective attenuation of the (a) outer and (b) middle ear (after [133]).

The transformed sound moves the eardrum, which, in turn, transfers the mechanical vibrations to the cochlea. The cochlear structure then induces traveling waves along the basilar membrane. The basilar membrane acts as a spectrum analyzer, spatially decomposing the signal into frequency components. A frequency-to-place transformation takes place along the length of the basilar membrane. More than 30,000 neural cells connect the basilar membrane to the auditory nerve that leads to the brain. These neural cells are tuned to different frequency bands. This means that each neural cell present on the basilar membrane would resonate to a particular set of frequencies. These frequency bands are typically measured in terms of the critical bandwidth (CB). A distance of one critical band is commonly referred to as "one Bark." The Bark scale [43], $z_b(f)$, is defined as,

$$z_b(f) = 13 \arctan (0.00076 f) + 3.5 \arctan \left[\left(\frac{f}{7500} \right)^2 \right] \tag{3.3}$$

where f is frequency in Hz. Other perceptually-relevant scales include the mel scale [150] and the equivalent rectangular bandwidth (ERB)-rate scale [70]. The mel scale [68, 150], $z_m(f)$, that is typically referred to as the pitch scale is popular in speech recognition.

$$z_m(f) = 1000 \log_2 \left(1 + \frac{f}{1000} \right) . \tag{3.4}$$

The mel scale was constructed as follows. In their experiments, Stevens and Volkman [151] chose 1000 Hz as equal to 1000 mels. Subjects were asked to change the frequency until the perceived pitch was twice, thrice, one-half, one-third, and so on. These pitch values were labeled as 2000 mels, 3000 mels, 500 mels, and 333 mels, respectively. Later a mapping between the linear Hz scale and the logarithmic-like mel scale was obtained [see Figure 3.5 (b)]. The mel scale only approximates the Bark scale and is easier to implement.

Early auditory filter shape experiments [44] suggested that the critical bandwidth below 500 Hz may change with center frequency. Moore and Glasberg [70] proposed an expression relating ERB-rate scale, $z_{ERB}(f)$, to center frequency.

$$z_{ERB}(f) = 11.7 \ln \left(\frac{f + 312}{f + 14675} \right) + 43 . \tag{3.5}$$

All the aforementioned scales are perceptually-significant and closely approximate the non-uniform frequency tiling of the auditory filter bank. Figure 3.5 shows the three scales against the frequency in Hertz scale. Relative to the Bark scale, the ERB scale yields a closer approximation to the psychophysical critical band assignment [69].

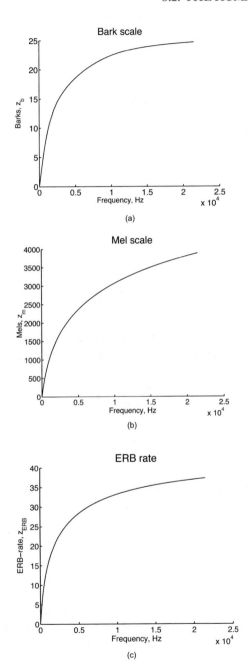

Figure 3.5: (a) The Bark scale, (b) the Mel scale, and (c) the ERB-rate scale.

```
%~~~~~~~~~~~~~~~~~~~~~~~~~~~~~~~~~~~~~~~~~~~~~~~~~~~~~~~~~~~~~~~~~~~~~~~~~~~~~~~~~~~~~~~~~~~
%% Bark scale, Mel scale, and ERB scale
%~~~~~~~~~~~~~~~~~~~~~~~~~~~~~~~~~~~~~~~~~~~~~~~~~~~~~~~~~~~~~~~~~~~~~~~~~~~~~~~~~~~~~~~~~~~
function [z_b] = Hz2Bark(f)

  % Hz to Bark scale
  fKhz = f/1000;
  z_b = 13 * atan(0.76*fKhz) + 3.5 * atan((fKhz/7.5).^2);

% Hz to Mel scale conversion
function [z_m] = Hz2Mel(f)

  % Hz to Mel scale
  fKhz = f/1000;
  z_m = 1000 * log2(1+fKhz);

% Hz to ERB scale conversion
function [z_erb] = Hz2ERB(f)

  z_erb = 11.7 * log((f+312)/(f+14675)) + 43;
```

Figure 3.6: The Bark scale, the Mel scale, and the ERB-rate scale.

The peripheral auditory system is typically modeled as a bank of non-uniform, overlapping bandpass filters. These auditory-filter bandwidths are approximated using either the CB or the ERB.

$$CB = 25 + 75 \left[1 + 1.4 \left(\frac{f}{1000} \right)^2 \right]^{0.69} \tag{3.6}$$

$$ERB = 24.7 \left[\frac{4.37 f}{1000} + 1 \right] . \tag{3.7}$$

A comparison of the CB and the ERB is shown in Figure 3.7. From this figure, the ERB scale implies that auditory filter bandwidths decrease below 500 Hz, whereas the CB remains essentially flat.

3.3   MASKING

A process where one sound is rendered inaudible because of the presence of another is called masking. If two sounds occur simultaneously and one is masked by the other, this is referred to as simultaneous masking. Similarly, a weak sound that precedes a louder sound is rendered inaudible. This effect is called forward temporal masking. The sound that is masked is called a maskee. These two scenarios are depicted in Figure 3.8. The presence of a strong noise or tone masker creates an excitation of sufficient strength on the basilar membrane to mask a weaker signal. Note that the y-axis in Figure 3.8 is represented as dB SPL. The sound pressure level (SPL) describes the intensity of sound pressure in dB relative to a standard reference level, 20 μN/m^2. This reference level is associated with the 0 dB

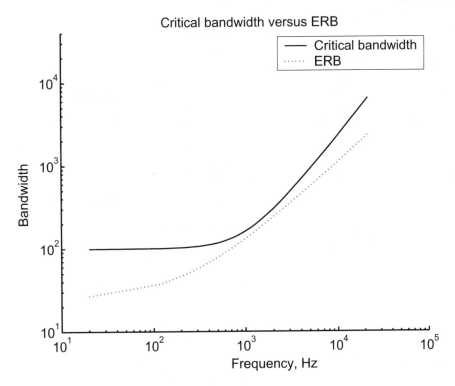

Figure 3.7: Comparison of the critical bandwidth (CB) and the equivalent rectangular bandwidth (ERB). The ERB implies that the auditory-filter bandwidth continues to decrease as center frequency decreases below 500 Hz.

SPL. In Figure 3.8 (a), the tone present at f_1 masks the tone at f_2. Similarly, the tone at f_3 masks the broadband noise. In Figure 3.8 (b), a masker appears at time t_1 and lasts until t_2. This masker created a sufficient temporal masked threshold that would make sounds, appearing just before and after, inaudible. Typically, the pre-masking lasts for around 10 ms, and post-masking duration can last upto 500 ms, depending on the temporal masked threshold.

3.3.1   MASKING ASYMMETRY

The ability of a sound to mask other sounds is determined by its tonality. For example, a tone requires a higher intensity to mask a noise-like maskee than a loud noise-like masker does to mask a tone. This is called the masking asymmetry. Painter and Spanias [47] presented an example that illustrates the asymmetry of masking (Figure 3.9). In Figure 3.9 (a), a 410 Hz pure tone of 76 dB PSL is masked by a narrowband noise of 80 dB SPL intensity centered at 410 Hz with CB 90 Hz. This type of simultaneous masking is called noise-masking-tone. Figure 3.9 (b) shows the tone-masking-noise

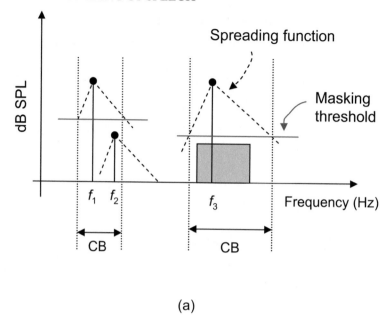

(a)

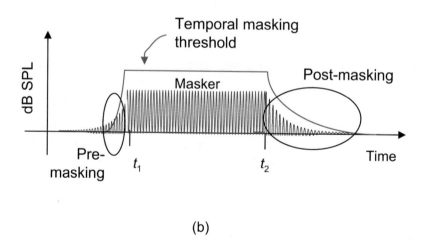

(b)

Figure 3.8: (a) Simultaneous or frequency masking, (b) Temporal masking. The presence of a strong noise or tone masker creates an excitation of sufficient strength on the basilar membrane to mask a weaker signal. In Figure 3.8 (a), the tone present at f_1 masks the tone at f_2. Similarly, the tone at f_3 masks the broadband noise. In Figure 3.8 (b), a masker appears at time t_1 and lasts until t_2. This masker created a sufficient temporal masked threshold that would make sounds, appearing just before and after, inaudible.

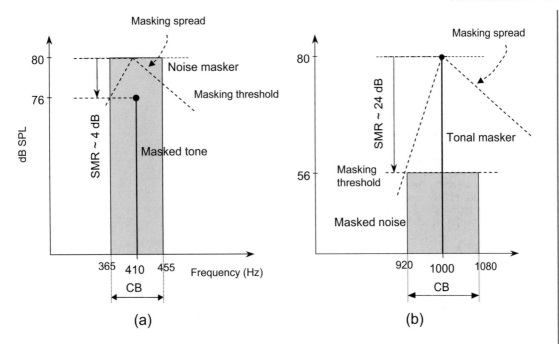

Figure 3.9: Greater masking power is associated with noise maskers than with tonal maskers. (a) Noise masking tone, and (b) Tone masking noise (after [47]). This figure illustrates that a tone requires a higher intensity to mask a noise-like maskee than a loud noise-like masker does to mask a tone.

scenario. A 1 kHz pure tone of 80 dB SPL intensity just masks a 56 dB SPL narrowband noise centered at 1 kHz with CB of 160 Hz. Note that the signal-to-mask ratio in case of noise-masking-tone is less than that of the tone-masking-noise case. Greater masking power is associated with noise maskers than with tonal maskers. The simultaneous masking effects described above (e.g., NMT, TMN) are not band-limited to within the boundaries of a single critical band. Inter-band masking where a masker centered within one critical band has some predictable effect on detection thresholds in other critical bands. This effect, also known as the spread of masking, is often modeled by an approximately triangular spreading function that has slopes of +25 and −10 dB per Bark. An analytical expression for spread of masking is given by,

$$SF(z_B) = 15.81 + 7.5(z_B + 0.474) - 17.5\sqrt{1 + (z_B + 0.474)^2}$$

where z_B is the Bark scale and $SF(z_B)$ is the masking spread in dB. In Figure 3.10, we present a Matlab program to experiment with tone-masking-noise and noise-masking tone scenarios. Experiment with different values of tone amplitude (alpha) and noise gain (beta) and find out when the broadband noise masks the tone and the tone masks the noise. More insight into masking principles can be

```
%~~~~~~~~~~~~~~~~~~~~~~~~~~~~~~~~~~~~~~~~~~~~~~~~~~~~~~~~~~~~~~~~~~~~~~~~~~~~~~~~~~~~~~~
% Tone-masking-noise and Noise-masking-tone experiment
%~~~~~~~~~~~~~~~~~~~~~~~~~~~~~~~~~~~~~~~~~~~~~~~~~~~~~~~~~~~~~~~~~~~~~~~~~~~~~~~~~~~~~~~
function [s1] = TMN_NMT_test()
  % Tone generator
    fc = 4000;        % fc in Hz
    alpha = 0.025;  % Tone amplitude
    len = 44099;   % length of the sinusoid
    fs = 44100;       % sampling rate
    % Generate a pure tone
    s1 = sin(2*pi*(fc/fs)*[1 :1 : len])';

  % Broadband noise generator
    beta = 1;
    s2 = [1, zeros(1, len-1)]';
    % Bandpass filter design
    [B, A] = butter(8, [2*3500/44100, 2*4500/44100]);
    s2 = filter(B, A, s2);

  % Test signal
    s = alpha * s1 + beta * s2;

%ASSIGNMENT:
% 1. For alpha = 0.025 and beta = 1, does the broad band noise completely mask the tone?
% 2. Experiment for different values of alpha and beta and find out when
%      the broadband noise masks the tone and the tone masks the noise.
% 3. For the TMN and NMT cases, plot the global masking threshold of the test signal and study
%      how the masking curve associated with noise falls below that of the tone for tone masking
%      noise case.
```

Figure 3.10: Asymmetry of masking.

gained by plotting the spread of masking (e.g., dotted line in Figure 3.9) using the Matlab programs given in Sections 3.4.1 and 3.4.2 for the MPEG psychoacoustic models 1 and 2.

3.4   PSYCHOACOUSTIC MODELS

In perceptual audio coders, typically, a global masking threshold curve is computed to estimate the just-noticeable distortion (JND) level introduced by quantization. Spectral components that fall below the JND level are considered to be perceptually-irrelevant and are usually not encoded. A psychoacoustic model is used to compute the global masking threshold. The ISO/IEC 11172-3 MPEG 1 standard [21] published two psychoacoustic models, i.e., model 1 and model 2. The psychoacoustic model 1 is computationally less complex than psychoacoustic model 2. In both the models, the main idea is to separate a complex sound into masker components, i.e., tonal and noise. Later, the properties of superposition of the masked thresholds are applied to these individual tonal/noise components to obtain the overall masked threshold. The psychoacoustic model 1 identifies tonal components based on the spectral peaks in a critical band. The remaining components within a critical band are lumped as one non-tonal component. The index of such a non-tonal component is chosen as the one that is closest to the geometric mean of all the components in that critical band. In the psychoacoustic model 2, the tonality decision is determined using a predictability measure. A polynomial predictor is used to predict the tonality index.

3.4.1 THE MPEG 1 PSYCHOACOUSTIC MODEL 1

The Psychoacoustic Model 1 (see flowchart Figure 3.11) employs the following steps to compute the global masking threshold:

1) Spectral analysis and SPL normalization,

2) Identification of tonal and noise maskers,

3) Decimation of the maskers to obtain only the perceptually-relevant maskers,

4) Calculation of the individual masking thresholds, and

5) Calculation of the global masking threshold in each band.

Figure 3.12 shows the Matlab implementation of psychoacoustic model 1 and different functions that implement the above five steps. The Matlab functions are further elaborated in Figure 3.13 through Figure 3.17.

3.4.1.1 Spectral Analysis
The spectral analysis procedure works as follows. First, incoming audio samples, $s(n)$, are normalized according to the FFT length, N, using the relation

$$x(n) = \frac{s(n)}{N} . \tag{3.8}$$

A power spectral density (PSD) estimate, $P(k)$, is then obtained using a 512-point FFT, i.e.,

$$P(k) = 90.3 + 10 \log_{10} \left| \sum_{n=0}^{N-1} w(n)x(n)e^{-j2\pi kn/N} \right|^2 \quad 0 \leq k \leq N/2 . \tag{3.9}$$

The Hann window, $w(n)$, is defined as

$$w(n) = \frac{1}{2} \left[1 - \cos\left(\frac{2\pi n}{N}\right) \right] . \tag{3.10}$$

3.4.1.2 Identification of Tonal and Noise Maskers
Local maxima in the sample PSD that exceed neighboring components within a certain Bark distance by at least 7 dB are classified as tonal. Specifically, the tonal set, S_T, is defined as

$$S_T = \left\{ P(k) \,\middle|\, \begin{array}{l} P(k) > P(k \pm 1), \\ P(k) > P(k \pm \Delta_k) + 7\, dB \end{array} \right\} \tag{3.11}$$

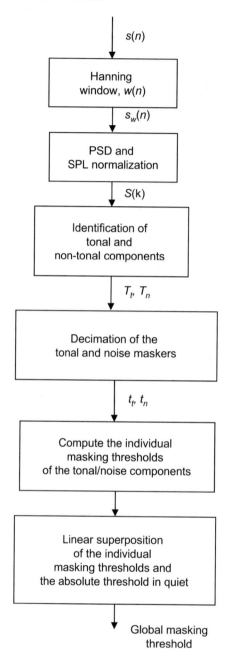

Figure 3.11: Flowchart depicting the steps involved in the ISO/IEC 11172-3 MPEG 1 Psychoacoustic Model 1.

```
% **********************************************************************
% ISO/IEC 11172-3 (MPEG-1) Psychoacoustic Model-1
% **********************************************************************
function [P, Thr_global] = psyModel_1(inFrame, frSize)

  % Define the constants
  fs = 44100;
  fftSize = frSize;  % FFT size

  % Freq. bins in Hz
  freq_hz = [1 : fftSize/2+1] * (fs / fftSize);
  % Bark indices corresponding to freq. bins
  freq_bark = 13 * atan(.00076*freq_hz) + 3.5 * atan((freq_hz/7500).^2);

  % Absolute Threshold in quiet
  Abs_thr = 3.64*(freq_hz/1000).^(-0.8) ...
            - 6.5*exp(-0.6*(freq_hz/1000-3.3).^2) ...
            + 0.001*(freq_hz/1000).^4;

  % Step-1: Spectral analysis and SPL Normalization
  [P] = psy_step_1(inFrame, fftSize);  % This function uses Eq. (3.8)-(3.10)

  % Step-2: Identification of tonal and noie maskers
  [P_TM, P_NM] = psy_step_2(P, freq_bark);  % This function uses Eq.(3.11) - (3.15)

  % Step-3: Decimation and re-organization of maskers
  [P_TM_th, P_NM_th] = psy_step_3(P_TM, P_NM, freq_bark, Abs_thr);  % Eq.(3.16) - (3.19)

  % Step-4: Calculation of individual masking thresholds
  [Thr_TM, Thr_NM] = psy_step_4(P_TM_th, P_NM_th, freq_bark);  % Eq.(3.20) - (3.22)

  % Step-5: Calculation of global masking thresholds
  [Thr_global] = psy_step_5(Thr_TM, Thr_NM, Abs_thr, freq_bark);  % Eq.(3.23)
```

Figure 3.12: Matlab implementation of the ISO/IEC 11172-3 MPEG 1 Psychoacoustic Model 1.

```
  %-------------------------------------------------------------------------
  % Step-1: Spectral analysis and SPL Normalization
  %-------------------------------------------------------------------------
  function [PSD] = psy_step_1(s, fftSize)

    % Power normalization term, PN
    PN = 90.3;
    % Normalize the input audio samples according to the FFT length
    %   and the number of bits per sample
    x = s/fftSize;
    %  Design the Hanning window, w(n)
    win = hanning(fftSize);
    %  Compute the power spectral density (PSD), P
    P = PN + 10*log10( (abs(fft(win.*x, fftSize))).^2 );
    PSD = P(1: fftSize/2+1);   % Only first half is required
```

Figure 3.13: Spectral analysis.

```
%-------------------------------------------------------------------------
% Step-2: Identification of tonal and noise maskers
%-------------------------------------------------------------------------
function [P_TM, P_NM] = psy_step_2(P, freq_bark)

  % Browse through the power spectral density, P, to determine the tone maskers
  P_TM = zeros(1, length(P));
  for k = 1 : length(P)
    if(tone_masker_check(P, k))
      % If index k corresponds to a tone,
      % combine the energy from three adjacent spectral components
      % centered at the peak to form a single tonal masker
      P_TM(k) = 10*log10(10.^(0.1.*P(k-1))+10.^(0.1.*P(k))+10.^(0.1.*P(k+1)));
    end
  end

  % Find noise maskers within the critical band
  P_NM = zeros(1, length(P_TM));
  % lower spectral line boundary of the critical bank, l
  lowbin = 1;
  % upper spectral line boundary of the critical bank, u
  highbin = max(find(freq_bark < 1));
  % loc is the geometric mean spectral line of the critical band,
  for band = 1:24
    [noise_masker_at_loc, loc] = noise_masker_check(P, P_TM, lowbin, highbin);
    if (loc ~= -1)
      P_NM(floor(loc)) = noise_masker_at_loc;
    end
    lowbin = highbin;
    highbin = max(find(freq_bark<(band+1)));
  end
  % Plotting
  figure(1), plot(freq_bark, P);
  hold on;
  hold on, plot(freq_bark, Abs_thr, 'r:');
  plot(freq_bark, P_TM, 'x');
  plot(freq_bark, P_NM, 'o');
  hold off,
  set(gca, 'ylim', [-20, 100]);
  legend('PSD', 'Absolute Threshold', 'Tone maskers', 'Noise maskers');
```

Figure 3.14: Identification of tonal and noise maskers. *Continues.*

```
%-------------------------------------------------------------------
function bool = tone_masker_check(P, k)
%-------------------------------------------------------------------

  % If P(k) is a local maxima and is greater than 7dB in
  % a frequency dependent neighborhood, it is a tone.
  % This neighborhood is defined as:
  %   within 2            if 2   < k < 63, for frequencies between 0.17-5.5kHz
  %   within 2,3             63 <= k < 127, for frequencies between 5.5-11Khz
  %   within 2,3,4,5,6    127 <= k < 256, for frequencies between 11-20Khz

  % If it is at the beginning or end of P, then it is not a local maxima
  % The if-else statements below computes the tonal set.
  if ((k<=1) | (k>=250))
    bool = 0;
  % if it's not a local maxima, leave with bool=0
  elseif ((P(k)<P(k-1)) | (P(k)<P(k+1)))
    bool = 0;
  % otherwise, we need to check if it is a max in its
  % neighborhood.
  elseif ((k>2) & (k<63))
    bool = ((P(k)>(P(k-2)+7)) & (P(k)>(P(k+2)+7)));
  elseif ((k>=63) & (k<127))
    bool = ((P(k)>(P(k-2)+7)) & (P(k)>(P(k+2)+7)) & (P(k)>(P(k-3)+7)) & (P(k)>(P(k+3)+7)));
  elseif ((k>=127) & (k<=256))
    bool = ((P(k)>(P(k-2)+7)) & (P(k)>(P(k+2)+7)) & (P(k)>(P(k-3)+7)) & (P(k)>(P(k+3)+7))
...
      & (P(k)>(P(k-4)+7)) & (P(k)>(P(k+4)+7)) & (P(k)>(P(k-5)+7)) & (P(k)>(P(k+5)+7)) ...
      & (P(k)>(P(k-6)+7)) & (P(k)>(P(k+6)+7)));
  else
    bool = 0;
  end
```

Figure 3.14: *Continued.* Identification of tonal and noise maskers. *Continues.*

where

$$\Delta_k \in \begin{cases} 2 & 2 < k < 63 & \text{(0.17-5.5 kHz)} \\ [2,3] & 63 \le k < 127 & \text{(5.5-11 kHz)} \\ [2,6] & 127 \le k \le 256 & \text{(11-20 kHz)} . \end{cases} \qquad (3.12)$$

Tonal maskers, $P_{TM}(k)$, are computed from the spectral peaks listed in S_T as follows

$$P_{TM}(k) = 10 \log_{10} \sum_{j=-1}^{1} 10^{0.1 P(k+j)} \text{ (dB)} . \qquad (3.13)$$

For each neighborhood maximum, energy from three adjacent spectral components centered at the peak are combined to form a single tonal masker. A single noise masker for each critical band, $P_{NM}(\bar{k})$, is then computed from remaining spectral lines not within the $\pm \Delta_k$ neighborhood of a

```
%-------------------------------------------------------------------
function [noise_masker_at_loc, loc] = noise_masker_check(psd, tone_masker, low, high)
%-------------------------------------------------------------------

   noise_members = ones(1,high-low+1);
   % Browse through the power spectral density, P, to determine the noise maskers
   for k = low:high
     % if there is a tone
     if (tone_masker(k) > 0)
       % check frequency location and determine neighborhood length
       if ((k>2) & (k<63))
         m = 2;
       elseif((k>=63) & (k<127))
         m = 3;
       elseif((k>=127) & (k<256))
         m = 6;
       else
         m = 0;
       end
       % Set all members of the neighborhood to 0, which
       % removes them from the list of noise members
       for n = (k-low+1)-m:(k-low+1)+m
         if (n > 0)
           noise_members(n) = 0;
         end
       end
     end
   end
   % If there are no noise members in the range, then leave
   if (isempty(find(noise_members)))
     noise_masker_at_loc = 0;
     loc = -1;
   else
     temp = 0;
     for k = (low+find(noise_members)-1)
       temp = temp + 10.^(0.1.*psd(k));
     end
     noise_masker_at_loc = 10*log10(temp);
     loc = geomean(low+find(noise_members)-1);
   end
```

Figure 3.14: *Continued.* Identification of tonal and noise maskers.

tonal masker using the sum

$$P_{NM}\left(\bar{k}\right) = 10\log_{10}\sum_{j} 10^{0.1P(j)} \text{ (dB)}, \forall P\left(j\right) \notin \{P_{TM}\left(k, k\pm1, k\pm\Delta_k\right)\} \tag{3.14}$$

```
%---------------------------------------------------------------
% Step-3: Decimation and re-organization of maskers
%---------------------------------------------------------------
function [TM_above_thres, NM_above_thres] = psy_step_3(tone_masker, noise_masker,
freq_bark, Abs_thr)

  TM_above_thres = tone_masker.*(tone_masker>Abs_thr);
  NM_above_thres = noise_masker.*(noise_masker>Abs_thr);
  % The remaining maskers must now be checked to see if any are within a critical band.
  % If they are, then only the strongest one matters.  The other can be set to zero.
  % Lets go through masker list.
  for j = 1 : length(Abs_thr)
    toneFound=0;
    noiseFound=0;
    % was a tone or noise masker found?
    if (TM_above_thres(j)>0)
      toneFound=1;
    end
    if (NM_above_thres(j)>0)
      noiseFound=1;
    end
    % if either masker found
    if (toneFound | noiseFound)
      masker_loc_barks = freq_bark(j);
      % determine low and high thresholds of critical band
      crit_bw_low  = masker_loc_barks-0.5;
      crit_bw_high = masker_loc_barks+0.5;
      % determine what indices these values correspond to
      low_loc  = max(find(freq_bark<crit_bw_low));
      if (isempty(low_loc))
        low_loc=1;
      else
        low_loc=low_loc+1;
      end
      high_loc = max(find(freq_bark<crit_bw_high));
```

Figure 3.15: Decimation of the maskers to obtain only the perceptually-relevant maskers. *Continues.*

where $\bar{k}$ is defined to be the geometric mean spectral line of the critical band, i.e.,

$$\bar{k} = \left(\prod_{j=l}^{u} j \right)^{1/(l-u+1)} \tag{3.15}$$

where l and u are the lower and upper spectral line boundaries of the critical band, respectively. In each critical band, Eq. (3.14) combines into a single noise masker all of the energy from spectral components that have not contributed to a tonal masker within the same band.

```
% At this point, we know the location of a masker and its critical band.
% Depending on which type of masker it is, browse through and eliminate
% the maskers within the critical band that are lower.
for k=low_loc:high_loc
  if (toneFound)
    % find other tone maskers in critical band
    if ((TM_above_thres(j) < TM_above_thres(k)) & (k ~= j))
      TM_above_thres(j)=0;
      break;
    elseif (k ~= j)
      TM_above_thres(k)=0;
    end
    % find noise maskers in critical band
    if (TM_above_thres(j) < NM_above_thres(k))
      TM_above_thres(j)=0;
      break;
    else
      NM_above_thres(k)=0;
    end
  elseif (noiseFound)
    % find other noise maskers in critical band
    if ((NM_above_thres(j) < NM_above_thres(k)) & (k ~= j))
      NM_above_thres(j)=0;
      break;
    elseif (k ~= j)
      NM_above_thres(k)=0;
    end
    % find tone maskers in critical band
    if (NM_above_thres(j) < TM_above_thres(k))
      NM_above_thres(j)=0;
      break;
    else
      TM_above_thres(k)=0;
    end
  else
    disp('Error in tone/noise masker locations.');
  end
  end % END of for k=low_loc:high_loc
  end    % END of if (toneFound | noiseFound)
end % END of for j = 1 : length(Abs_thr)
% Plotting
figure(2), plot(freq_bark, P);
hold on, plot(freq_bark, Abs_thr, ':');
plot(freq_bark, P_TM_th, 'rx');
plot(freq_bark, P_NM_th, 'ko');
set(gca, 'ylim', [-20, 100]);
legend('PSD', 'Absolute Threshold', 'Selected tone maskers', 'Selected noise maskers');
hold off,
```

Figure 3.15: *Continued.* Decimation of the maskers to obtain only the perceptually-relevant maskers.

```
%-----------------------------------------------------------------------
% Step-4: Calculation of individual masking thresholds
%-----------------------------------------------------------------------
function [Thr_TM, Thr_NM] = psy_step_4(P_TM_th, P_NM_th, freq_bark)

   Thr_TM = zeros(1,length(P_TM_th));
   % Go through the tone list
   for k = find(P_TM_th),
     % determine the masking threshold around the tone masker
     % CALL function 5: mask_threshold
     [thres, start] = mask_threshold(1,k,P_TM_th(k),freq_bark);
     % Add the power of the threshold to temp in the proper frequency range
     Thr_TM(start:start+length(thres)-1)=Thr_TM(start:start+length(thres)-
1)+10.^(0.1.*thres);
   end
   %   figure, plot(THR)
   Thr_NM = zeros(1,length(P_NM_th));
   % Go through noise list
   for k = find(P_NM_th)
     % determine the masking threshold around the noise masker
     % CALL function 5: mask_threshold
     [thres, start] = mask_threshold(0,k,P_NM_th(k),freq_bark);
     % add the power of the threshold to temp in the proper frequency range
     Thr_NM(start:start+length(thres)-1)=Thr_NM(start:start+length(thres)-
1)+10.^(0.1.*thres);
   end
   % Plotting
   figure(3), plot(freq_bark, 10*log10(Thr_TM+eps), 'r:');
   hold on, plot(freq_bark, 10*log10(Thr_NM+eps), 'k:');
   legend('Tone masking threshold', 'Noise masking threshold');
   set(gca, 'ylim', [-20, 100]);
   hold off,
```

Figure 3.16: Calculation of the individual masking thresholds. *Continues.*

3.4.1.3 Decimation of the Maskers

Any tonal or noise maskers below the absolute threshold are discarded, i.e., only maskers which satisfy

$$P_{TM,NM}(k) \geq T_q(k) \tag{3.16}$$

are retained, where $T_q(k)$ is the SPL of the threshold in quiet at spectral line k. Next, a sliding 0.5 Bark-wide window is used to replace any pair of maskers occurring within a distance of 0.5 Bark by the stronger of the two. After the sliding window procedure, masker frequency bins are reorganized according to the subsampling scheme

$$P_{TM,NM}(i) = P_{TM,NM}(k) \tag{3.17}$$

$$P_{TM,NM}(k) = 0 \tag{3.18}$$

```
%---------------------------------------------------------------
function [threshold, start] = mask_threshold(type, j, P, bark)
%---------------------------------------------------------------
  % Determine where masker is in barks
  maskerloc = bark(j);
  % set up range of the resulting function in barks
  low = maskerloc - 3;
  high = maskerloc + 8;
  % In discrete bins
  lowbin = max(find(bark<low));
  if (isempty(lowbin))
    lowbin = 1;
  end
  highbin=max(find(bark<high));
  % calculate spreading function
  SF = spreading_function(j, P, lowbin, highbin, bark);
  if (type==0)
    % calculate noise threshold
    threshold = P-0.175*bark(j)+SF-2.025;
  else
    % calculate tone threshold
    threshold = P-0.275*bark(j)+SF-6.025;
  end
  % The lowest value in threshold corresponds to the frequency bin at lowbin.
  start = lowbin;

%---------------------------------------------------------------
function spread = spreading_function(masker_bin, power, low, high, bark)
%---------------------------------------------------------------

  masker_bark=bark(masker_bin);
  for i=low:high,
    maskee_bark=bark(i);
    deltaz=maskee_bark-masker_bark;
    if ((deltaz>=-3.5) & (deltaz<-1))
      spread(i-low+1)=17*deltaz-0.4*power+11;
    elseif ((deltaz>=-1) & (deltaz<0))
      spread(i-low+1)=(0.4*power+6)*deltaz;
    elseif ((deltaz>=0) & (deltaz<1))
      spread(i-low+1)=-17*deltaz;
    elseif ((deltaz>=1) & (deltaz<8.5))
      spread(i-low+1)=(0.15*power-17)*deltaz-0.15*power;
    end
  end
```

Figure 3.16: *Continued.* Calculation of the individual masking thresholds.

```
%-----------------------------------------------------------------------
% Step-5: Calculation of global masking thresholds
%-----------------------------------------------------------------------
function [Thr_global] = psy_step_5(Thr_TM, Thr_NM, Abs_thr, freq_bark)

  % Thr_TM: Individual TMN masking threshold from Section 3.4.1.4
  % Thr_NM: Individual NMT masking threshold from Section 3.4.1.4
  % Abs_thr: Absolute Threshold in quiet from Matlab program, Figure 3.12

  % Estimate the linear superposition of TMN and NMT components.
  Thr_TM_NM = Thr_TM + Thr_NM;

  % Add the masking threshold contribution from the absolute hearing threshold
  Thr_global = Thr_TM_NM + 10.^(0.1.*Abs_thr);

  % Plotting
  figure(4), plot(freq_bark, 10*log10(Thr_global+eps), 'k');
  legend('JND curve');    % The global masking threshold or Just noticeable distortion
  set(gca, 'ylim', [-20, 100]);
  hold off,
```

Figure 3.17: Calculation of the global masking threshold.

where

$$
i = \begin{cases}
k & 1 \leq k \leq 48 \\
k + (k \bmod 2) & 49 \leq k \leq 96 \\
k + 3 - ((k-1) \bmod 4) & 97 \leq k \leq 232 .
\end{cases}
\tag{3.19}
$$

The net effect of Eq. (3.19) is 2:1 decimation of masker bins in critical bands 18-22 and 4:1 decimation of masker bins in critical bands 22-25, with no loss of masking components. This procedure reduces the total number of tone and noise masker frequency bins under consideration from 256 to 106.

3.4.1.4 Individual Masking Thresholds

Using the decimated set of tonal and noise maskers, individual tone and noise masking thresholds are computed next. Each individual threshold represents a masking contribution at frequency bin i due to the tone or noise masker located at bin j. Tonal masker thresholds, $T_{TM}(i, j)$, are given by

$$
T_{TM}(i, j) = P_{TM}(j) - 0.275z(j) + SF(i, j) - 6.025 \quad \text{(dB SPL)}
\tag{3.20}
$$

where $P_{TM}(j)$ denotes the SPL of the tonal masker in frequency bin j, $z(j)$ denotes the Bark frequency of bin j, and the spread of masking from masker bin j to maskee bin i, $SF(i, j)$, is modeled by the expression

$$SF(i,j) = \begin{cases} 17\Delta_z - 0.4P_{TM}(j) + 11, & -3 \le \Delta_z < -1 \\ (0.4P_{TM}(j) + 6)\Delta_z, & -1 \le \Delta_z < 0 \\ -17\Delta_z, & 0 \le \Delta_z < 1 \\ (0.15P(j)_{TM} - 17)\Delta_z - 0.15P(j)_{TM}, & 1 \le \Delta_z < 8 \end{cases} \quad \text{(dB SPL)} \quad (3.21)$$

i.e., as a piecewise linear function of masker level, $P(j)$, and Bark maskee-masker separation, $\Delta_z = z(i) - z(j)$. Individual noise masker thresholds, $T_{NM}(i,j)$, are given by

$$T_{NM}(i,j) = P_{NM}(j) - 0.175z(j) + SF(i,j) - 2.025 \quad \text{(dB SPL)} \quad (3.22)$$

where $P_{NM}(j)$ denotes the SPL of the noise masker in frequency bin j, $z(j)$ denotes the Bark frequency of bin j, and $SF(i,j)$ is obtained by replacing $P_{TM}(j)$ with $P_{NM}(j)$ everywhere in Eq. (3.21).

3.4.1.5 Global Masking Threshold

In this step, individual masking thresholds are combined to estimate a global masking threshold for each frequency bin in the subset given by Eq. (3.19). The model assumes that masking effects are additive. The global masking threshold, $T_g(i)$, is therefore obtained by computing the sum

$$T_g(i) = 10\log_{10}\left(10^{0.1T_q(i)} + \sum_{l=1}^{L} 10^{0.1T_{TM}(i,l)} + \sum_{m=1}^{M} 10^{0.1T_{NM}(i,m)}\right) \quad \text{(dB SPL)} \quad (3.23)$$

where $T_q(i)$ is the absolute hearing threshold for frequency bin i, $T_{TM}(i,l)$ and $T_{NM}(i,m)$ are the individual masking thresholds, and L and M are the numbers of tonal and noise maskers, respectively. Figure 3.18 presents an example simulation of the ISO/IEC11172-3 MPEG 1 psychoacoustic model 1.

Figure 3.18 (a) shows the time-domain plot of an audio segment. The PSD and the absolute threshold in quiet are shown in Figure 3.18 (b). The identified tonal ("x") and non-tonal ("o") components are shown in Figure 3.18 (c). The individual masked thresholds of the tonal and non-tonal components after decimation are shown in Figure 3.18 (e). A global masking threshold, Figure 3.18 (f), is then obtained by combining these individual masked thresholds with the absolute threshold in quiet. After the global masking threshold is estimated, the minimum masking threshold within a critical band is calculated. From Figure 3.18 (f), it is evident that some portions of the input spectrum require SNRs of better than 30 dB to prevent audible distortion, while other spectral regions require less than 5 dB SNR. In fact, some high-frequency portions of the signal spectrum are masked (see signal spectrum falling below the JND curve) and therefore are perceptually irrelevant, ultimately requiring no bits for quantization without the introduction of artifacts.

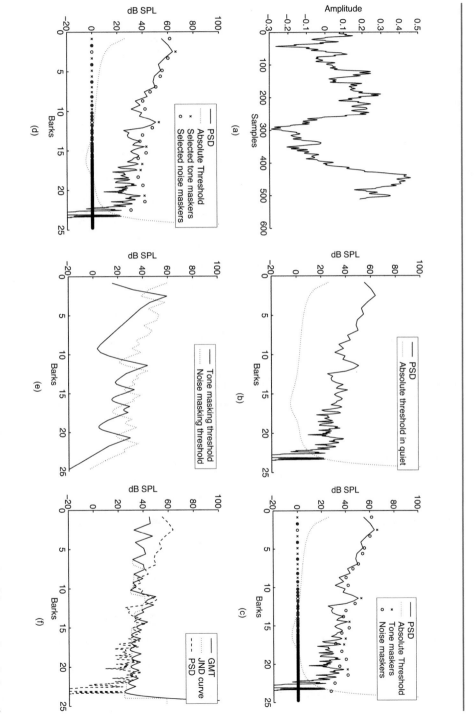

Figure 3.18: The ISO/IEC MPEG 11172-3 MPEG 1 psychoacoustic model 1 simulation. (a) Input audio, (b) spectral analysis and SPL normalization, (c) identification of tonal and noise maskers, (d) decimation of the maskers, (e) individual masking thresholds, and (f) global masking threshold (GMT).

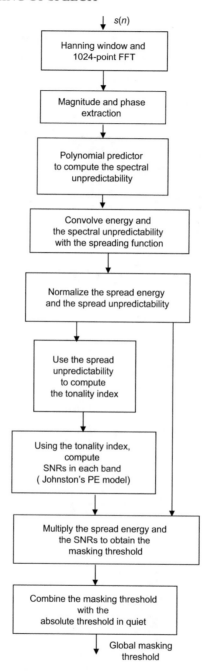

Figure 3.19: Flowchart depicting the steps involved in the ISO/IEC 11172-3 MPEG 1 Psychoacoustic Model 2.

3.4.2   THE MPEG 1 PSYCHOACOUSTIC MODEL 2

The Psychoacoustic Model 2 (see flowchart Figure 3.19) employs the following steps to compute the global masking threshold:

1) spectral analysis,

2) extract magnitude and phase,

3) generate magnitude and phase predictions,

4) compute spectral unpredictability measure [using Eq. (3.24)],

5) compute energy and unpredictability in the threshold partitions,

6) convolve both the partitioned energy and the unpredictability with the spreading function,

7) generate tonality index from the normalized spread unpredictability [using Eq. (3.25)],

8) calculate the required SNR in each partition using the tonality index. This step is similar to Johnston's perceptual entropy [152, 153] calculation,

9) multiply the normalized spread energy with the required SNRs to obtain the energy threshold, and

10) add absolute threshold in quiet to the energy threshold to obtain the global masking threshold.

Figure 3.20 shows the Matlab implementation of psychoacoustic model 2 and different functions that implement the above steps.

Figure 3.21 presents an example simulation of the ISO/IEC 11172-3 MPEG 1 psychoacoustic model 2. Figure 3.21 (a) shows the time-domain plot of an audio segment. The PSD and the absolute threshold in quiet are shown in Figure 3.21 (b). A 1024-point FFT is used to obtain the magnitude-phase (r, θ) representation of the audio segment. A simple polynomial predictor is used to estimate the magnitude and phase $(\hat{r}, \hat{\theta})$ from the previous two frames (see Step 3 in Matlab Program). The spectral unpredictability measure is computed as follows:

$$
w_{sp} = \frac{\sqrt{\left[r\cos(\theta) - \hat{r}\cos(\hat{\theta})\right]^2 + \left[r\sin(\theta) - \hat{r}\sin(\hat{\theta})\right]^2}}{r + \hat{r}} . \tag{3.24}
$$

Figure 3.21 (c) shows the energy (i.e., r^2) and the spectral unpredictability, w_{sp}. The next step is to convolve the energy and the spectral unpredictability with a triangular spreading function [e.g., Figure 3.21 (d)]. The resulting spread energy and the spread unpredictability are then normalized to compensate for energy weighting. Figure 3.21 (e) shows the normalized spread energy, ε. The normalized spread unpredictability, w_{Nsp}, is used to compute the tonality index.

$$
t_b = -0.4343\log\left(w_{Nsp}\right) - 0.30103 . \tag{3.25}
$$

```matlab
% ***********************************************************************
% ISO/IEC 11172-3 (MPEG-1) Psychoacoustic Model-2
% ***********************************************************************
function [P, Thr_global] = psyModel_2(inFrame, frSize)

    % Define the constants
    fs = 44100;
    fftSize = frSize;  % FFT size
    % Freq. bins in Hz
    freq_hz = [1 : fftSize/2+1] * (fs / fftSize);
    % Bark indices corresponding to freq. bins
    freq_bark = 13 * atan(.00076*freq_hz) + 3.5 * atan((freq_hz/7500).^2);
    %%% Set up the Bark scale low and upper bin indices for partitions %%%
    % Number of threshold partitions.
    numPartitions = 64;
    % Bark step
    barkStep = (freq_bark(end) - freq_bark(1))/(numPartitions);
    barkPartition = [freq_bark(1) : barkStep : freq_bark(end)];
    f = interp1(freq_bark, freq_hz, barkPartition, 'spline');
    barkIndices = ceil(f*fftSize/fs);
    % The lower and upper frequency bin number for each Bark band.
    barkFreq = [barkIndices(1:end-1).', barkIndices(2:end).'-1];
    % Adjust the upper frequency bin associated with the last Bark band.
    barkFreq(numPartitions, 2) = fftSize/2 + 1;
    b_lower = barkFreq(:,1);
    b_upper = barkFreq(:,2);

    freq_median = mean(barkFreq.') * (fs/fftSize);
    b_median = 13 * atan(.00076*freq_median) + 3.5 * atan((freq_median/7500).^2);

    % Initializations (magnitude and phase of spectrum)
    r1 = 0; r2 = 0; theta1 = 0; theta2 = 0;

    % Spreading function, SPF
    SPF = zeros(numPartitions, numPartitions);
    for ic = 1 : numPartitions
      for jc = 1 : numPartitions
        x = 1.05*(b_median(jc) - b_median(ic)) - 0.5;
        tx = 8 * min((x^2 - 2*x), 0);
        y = 1.05*(b_median(jc) - b_median(ic)) + 0.474;
        ty = 15.811389 + 7.5*y - 17.5*sqrt(1+y^2);
        if ty < -100
          SPF(ic,jc)=0;
        else
          SPF(ic,jc)=10^((tx+ty)/10);
        end
      end
    end
```

Figure 3.20: Matlab implementation of the ISO/IEC 11172-3 MPEG 1 Psychoacoustic Model 2. *Continues.*

```
% SPF normalization constants for each threshold partition
for ic = 1 : numPartitions
  SPF_norm(ic) = 1/sum(SPF(1:numPartitions, ic));
end

% Absolute Threshold in quiet
Abs_thr = 3.64*(freq_hz/1000).^(-0.8) - 6.5*exp(-0.6*(freq_hz/1000-3.3).^2) ...
          + 0.001*(freq_hz/1000).^4;

% Step-1: Spectral analysis
win = hanning(fftSize);
S = fft(win.*(inFrame/fftSize), fftSize);
S = S(1: fftSize/2+1);    % Only first half is required

% Step-2: Extract magnitude and phase
r = abs(S);
theta = angle(S);

% Step-3: Generate magnitude and phase predictions
rHat  = 2*r1 - r2;  % r1 and r2 are magnitudes from past two frames
thHat = 2*theta1 - theta2;  % theta1 and theta2 are phase responses from past two frames

% Step-4: Compute spectral unpredictability measure [using Eq. (3.24)]
numer1 = r.*cos(theta) - rHat.*cos(thHat);
numer2 = r.*sin(theta) - rHat.*sin(thHat);
denom = r + rHat + eps;
spUnpred = sqrt(numer1.^2 + numer2.^2) ./ (denom);

% Step-5: Compute energy and unpredictability in the threshold partitions
for ic = 1 : numPartitions
  ib = [b_lower(ic) : b_upper(ic)];
  ener(b) = sum(r(ib).*r(ib));
  spUnpredEner(b) = sum(r(ib).*r(ib).*spUnpred(ib));
end

% Step-6: Convolve both the partitioned energy and the unpredictability
% with the spreading function (Figure 3.19(d)
for ic = 1 : numPartitions
  enerConv(ic) = sum(ener(1:numPartitions).*SPF(1:numPartitions, ic));
  spUnpredEnerConv(ic) = sum(spUnpredEner(1:numPartitions).*SPF(1:numPartitions,ic));
end
normUnpred = spUnpredEnerConv./(enerConv+eps); % Normalized partition unpredictability
normEner = enerConv .* SPF_norm;           % Normalized partition energy
```

Figure 3.20: *Continued.* Matlab implementation of the ISO/IEC 11172-3 MPEG 1 Psychoacoustic Model 2. *Continues.*

```
% Step-7: Generate tonality index from the normalized spread unpredictability [using Eq.
(3.25)],
  ic = find(normUnpred > 0.5);
  normUnpred(ic) = 0.5;
  ic = find(normUnpred < 0.05);
  normUnpred(ic) = 0.05;
  toneIndex = -0.434294482 * log(normUnpred.') - 0.301029996;

% Step-8: calculate the required SNR in each partition using the tonality index.
% This step is similar to Johnston's perceptual entropy [152] [153] calculation.
  NMT = 5.5;
  snrCutoff = [20*ones(1,11), 17, 17, 17, 15, 15, 10, 10, 7, 4.4, 4.4, 4.5*ones(38),
3.5*ones(1,4)];
  TMN = [24.5, 24.5, 24.5, 24.5, 24.5, 24.5, 24.5, 24.5, 24.5, 24.5, 24.5, ...
         24.5, 24.5, 24.5, 24.5, 24.5, 24.5, 24.5, 24.5, 24.5, 24.59, 25.05, ...
         25.5, 25.91, 26.32, 26.7, 27.06, 27.42, 27.82, 28.28, 28.71, 29.11, 29.51, ...
         29.85, 30.19, 30.57, 30.96, 31.34, 31.72, 32.12, 32.5, 32.84, 33.21, 33.59, ...
         33.92, 34.26, 34.61, 34.97, 35.32, 35.66, 35.99, 36.34, 36.68, 37.02, 37.37, ...
         37.71, 38.05, 38.4, 38.73, 39.06, 39.43, 39.73, 40.04];
  snr_dB = max([snrCutoff,(toneIndex.*TMN)+((1-toneIndex).*NMT)]);

% Step-9: Multiply the normalized spread energy with the required SNRs to obtain the
energy threshold
  snr = 10.^(-snr_dB./10.0);
  nb = normEner .* snr;

% Step-10: Add absolute threshold in quiet to the energy threshold to obtain the global
masking threshold.
  for ic = 1 : numPartitions
    ib = [b_lower(ic) : b_upper(ic)];
    nw(ib) = 10*log10(nb(ic)/length(ib)) + 96;
  end
  thr=max([nw, Abs_thr]);

% Update magnitude and phase buffers (2 = 2 previous, 1 = previous) prediction memory
  r2 = r1; r1 = r;
  theta2 = theta1; theta1 = f;
```

Figure 3.20: *Continued.* Matlab implementation of the ISO/IEC 11172-3 MPEG 1 Psychoacoustic Model 2.

After estimating the tonality index, t_b, Johnston's perceptual entropy method [152] is used to compute the required SNR in each critical band. An estimate of the energy threshold [Figure 3.21 (e)] is then obtained by multiplying the normalized spread energy, ε, with the SNR. A global masking threshold (Figure 3.21 (f)) is computed by adding the contribution from the absolute threshold of hearing to the energy threshold.

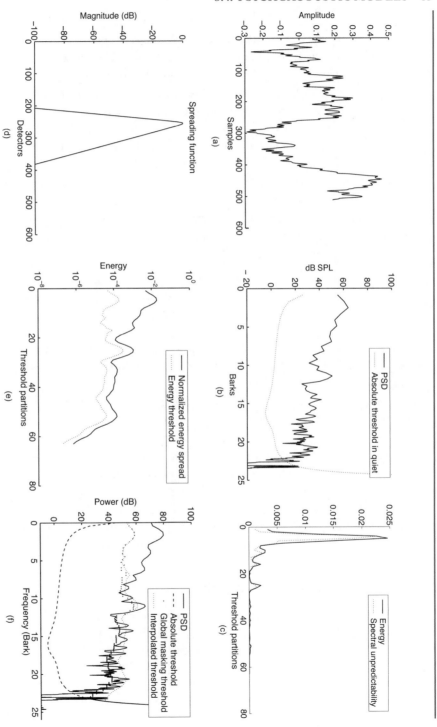

Figure 3.21: The ISO/IEC MPEG 11172-3 MPEG 1 psychoacoustic model 2 simulation. (a) Input audio, (b) spectral analysis and SPL normalization, (c) energy and spectral unpredictability in the threshold partitions, (d) spreading function; the x-axis is an equi-Bark-spaced set of 512 detectors, (e) normalized spread energy, and (f) global masking threshold.

3.4.3 A NON-LINEAR PSYCHOACOUSTIC MODEL

Most psychoacoustic models use a tonality estimation to obtain the masked threshold. In low-delay audio coding, it is customary to employ shorter frames. This means that the psychoacoustic model operates on shorter blocks of data leading to low-resolution frequency analysis. A reliable tonality estimation is more difficult when a low resolution (less than 128-point FFT) spectral analysis is employed. To this end, recently, several models have been proposed [147, 148] [154] to improve the estimate of the global masked threshold.

From psychoacoustic measurements, it is well-known that linear combination of individual masked thresholds often results in a lower overall masked threshold than determined experimentally. To this end, Baumgarte et al. [154] proposed a psychoacoustic model that is based on the nonlinear superposition of the masking components. The idea for the nonlinear superposition came from the model proposed by Lutfi [155]. After the input audio is decomposed to tonal and non-tonal maskers and the associated individual masked thresholds, $T_k(z_b)$ were computed, the following nonlinear superposition is used to compute the global masked threshold,

$$T(z_b) = \left[\sum_k (T_k(z_b))^\alpha \right]^{1/\alpha} \tag{3.26}$$

where $T_k(z_b)$ represents the individual masked threshold of the k-th masker, and α is the nonlinear superposition factor. Typically, $\alpha=0.3$ is used. Note that for $\alpha=1$, the above model simplifies to linear superposition (e.g., Figure 3.17).

The use of nonlinear superposition in a psychoacoustic model offers an improved approximation of the masked threshold evoked by the complex stimuli. This yields more accurate signal-to-mask ratios resulting in an improved perceptual bit allocation performance. The uniform spectral decomposition employed at the frontend of a psychoacoustic model does not align well with the non-uniform cochlear filter bank structure. To this end, a psychoacoustic model that closely mimics the non-uniform analysis of the cochlear filter bank structure was introduced by Baumgarte [147]. Van de Par et al. [148] proposed a psychoacoustic model where masking distortions are computed based on a spectral integration rule that accounts for distortions in adjacent auditory filters also. In this model, the basilar membrane was approximated using a gamma-tone filter bank.

3.5 AEP COMPUTATION

An auditory excitation pattern describes the effective energy spectrum reaching the cochlea. Several techniques are available to compute the AEP [132, 133], [43, 44]. These methods are based on 1) the auditory filter shapes and 2) the parametric spreading function. In terms of the filter bank analogy, the AEP can be thought of as the output of each auditory filter as a function of filter center frequency.

```
%--------------------------------------------------------------------------
% Non-linear psychoacoustic model example
%--------------------------------------------------------------------------
function [Thr_global] = nonlinear_GMT(Thr_TM, Thr_NM, Abs_thr)

    % Thr_TM: Individual TMN masking threshold from Section 3.4.1.4
    % Thr_NM: Individual NMT masking threshold from Section 3.4.1.4
    % Abs_thr: Absolute Threshold in quiet from Matlab program, Figure 3.12

    % alpha, non-linear superposition factor
    alpha = 0.3;

    % Estimate the non-linear superposition of TMN and NMT components.
    Thr_TM_NM = (Thr_TM.^alpha + Thr_NM.^alpha).^(1/alpha);

    % Add absolute hearing threshold
    Thr_global = Thr_TM_NM + 10.^(0.1.*Abs_thr);

    % Plotting
    figure, plot(freq_bark, 10*log10(Thr_global+eps), 'k');
    legend('JND curve');    % The global masking threshold or Just noticeable distortion
    set(gca, 'ylim', [-20, 100]);
    hold off,
```

Figure 3.22: Global masking threshold using non-linear superposition of TMN and NMT.

3.5.1 AUDITORY FILTER SHAPE METHOD

Figure 3.23 illustrates the steps involved in AEP computation using the auditory filter shape method. A 1.5 kHz sinusoid is used as the test stimulus. First, auditory filter shapes for several center frequencies, e.g., f_c = 750 Hz, 1000 Hz, 1500 Hz, 2000 Hz, 3000 Hz, and 4000 Hz are obtained. The auditory filter shapes are computed using either the notch-noise or ripple-noise experiments [43]. A model for the filter shape was derived [70] from these experiments, i.e.,

$$W(\Delta_f) = \left(1 + p\Delta_f\right)e^{-p\Delta_f} \tag{3.27}$$

where $\Delta_f = (f_c - f)/f_c$ is the deviation in frequency, f, from the auditory filter center frequency divided by the center frequency. The parameter, $p = 4f_c/$ERB, determines the shape of the passband of the filter. The ERB is estimated using Eq. (3.7). The above fitting function for the auditory filter shape is called the "rounded-exponential" model.

Next, the points where the 1.5 kHz tone intersects these auditory filter shapes are marked as a, b, c, and so on [see Figure 3.23 (a)]. The excitation pattern of a 1.5 kHz tone is then obtained by plotting the output of each filter as a function of filter center frequency. Joining these output points a-d-f-e-c-b will give us the excitation pattern, Figure 3.23 (b). A spline-fit that preserves the shape of the pattern is used for smooth interpolation. It is important to note that although the filter shapes are symmetric on a linear frequency scale, the AEP obtained is asymmetric around the center

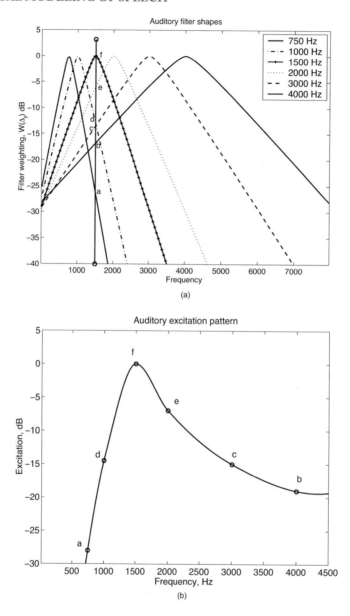

Figure 3.23: AEP computation from auditory filter shapes. (a) Simplified auditory filter shapes at several center frequencies, e.g., at 750 Hz, 1000 Hz, 1500 Hz, 2000 Hz, 3000 Hz, and 4000 Hz. Equation (3.27) is used to simulate these filter shapes. (b) The excitation pattern of a 1.5 kHz tone. The AEP is obtained by calculating the output of each filter as a function of filter center frequency. Joining these output points a-d-f-e-c-b gives us the excitation pattern. A spline-fit is used here for smooth interpolation.

frequency, i.e., f_c=1.5 kHz. For complex stimuli, AEPs from the individual sinusoidal components are added (in the linear power scale) to obtain the overall excitation pattern.

AEP computed using the auditory filer shape method suffers from resolution problems. Note that in the above simulation, we used only six auditory filter outputs (points "a"–"f") to generate the AEP. Usually, this will not be the case. In order to obtain AEPs of sufficient resolution, it was recommended [70] that filter center frequencies, f_c, must be chosen every 10 Hz. For the above simulation, this would result in approximately 325 points, which leads to a high complexity particularly when the signal contains complex stimuli.

```
%-----------------------------------------------------------------------------
% Auditory Excitation Pattern estimation using parametric spreading function method
%-----------------------------------------------------------------------------
function [AEP] = Excitation_Patt(s, fs, fft_size)

  % Freq. bins in Hz
  freq_hz = [1:fft_size/2+1] * (fs/fft_size);

  % Initialize soundfield-to-eardrum
  earComp = 6.5.*exp(-0.6*(((freq_hz/1000)-3.3).^2)) - 1e-3.*((freq_hz/1000).^4);

  % Initialize spreading function
  SF = computeSF(fs);

  % Generate mapping vector from Hz to localized Basilar energy distribution
  W = computeBasilarMapping(fft_size, fs);

  % Spectral analysis and SPL Normalization
  P = spectralAnalysis(s, fft_size);
  % Apply soundfield-to-eardrum and middle ear corrections
  Pc = P + earComp;
  % Return to linear scale
  Pc = 10.^(Pc./10);

  % Transform DFT components into a localized Basilar energy distribution
  PLoc = Pc * W;

  % Convolve localized Basilar energy distribution with the prototype spreading function
  Pspread = SpreadBasilar(PLoc.', SF);

  % Normalize spread version
  Gs = sum(Pspread)/sum(PLoc);
  Pspread = (1/Gs) * Pspread;

  % Auditory excitation pattern
  AEP = 10*log10(Pspread);
```

Figure 3.24: AEP computation using auditory filter shape method.

3.5.2   PARAMETRIC SPREADING FUNCTION METHOD

AEP computation, using parametric spreading function [132, 133], became popular in the early 1990s. The method (see Figure 3.25) works as follows. First, the input signal, $s(n)$, is mapped to the spectral domain, $S(k)$, using a windowed FFT. The Hanning window is typically used, i.e.,

$$w_{\text{Hann}}(n) = 0.5 \left[1 - \cos \left(\frac{2\pi n}{N} \right) \right] \tag{3.28}$$

where is the window length. Spectral components are then referenced to an assumed playback sound pressure level (SPL), such that, a full-scale sinusoid when perfectly resolved corresponds to 96 dB SPL. A spectral weighting function is next applied that emulates the filtering characteristics of the outer and middle ear. The ear transfer functions (Figure 3.4) are represented in terms of their associated frequency responses [133].

Next, the modified spectral components are mapped from the Hertz scale to a localized Bark energy distribution, $\mathbf{S}_l$, on an equi-Bark-spaced set of 512 discrete detectors.

$$z_b(f) = 13 \arctan (0.00076 f) + 3.5 \arctan \left[\left(\frac{f}{7500} \right)^2 \right]. \tag{3.29}$$

The physiological motivation for this equi-Bark-spaced detector set is derived from the spacing of neural hair cell detectors along the basilar membrane [132]. Note that the ERB-rate scale, Eq. (3.5) can also be used as suggested in [133] instead of the Bark scale. The localized detector energy is then smeared according to a parametric spreading function. This is given by,

$$\mathbf{S}_s = \left(\frac{l - u}{2} \right) (\mathbf{S}_l + c) - \left(\frac{l + u}{2} \right) \sqrt{t + (\mathbf{S}_l + c)^2} \tag{3.30}$$

where l=27 and u=10 were chosen for the lower and upper slopes (in dB per Bark) of the spreading function, t=0.1 and c=0.16. The parameter, t, provides desired smoothness (rounded) at the peaks. This smeared detector energy in dB SPL corresponds to the auditory excitation pattern, E_p.

Figure 3.26 (a)-(h) show the plots of intermediate signals in the AEP computation. A 512-sample, 1.5 kHz sinusoid, sampled at 16 kHz is used as the test signal. The Hanning windowed sinusoid, $s_w(n)$, is shown in Figure 3.26 (b). The PSD is plotted in Figure 3.26 (c). The modified PSD after the ear correction is shown in Figure 3.26 (d). The modified PSD is mapped from the linear Hz scale onto an equi-Bark-spaced scale. The resulting localized Bark energy distribution, $\mathbf{S}_l$, is shown in Figure 3.26 (e). The parametric spreading function, $\mathbf{T}_{sf}$, shown in Figure 3.26 (f) is convolved with the Bark energy distribution, $\mathbf{S}_l$, to obtain the smeared detector energy, $\mathbf{S}_s$. Figure 3.26 (h) shows the final excitation pattern, E_p, of the 1.5 kHz sinusoid.

3.6   PERCEPTUAL LOUDNESS

Figure 3.28 shows the steps involved in loudness computation [133].

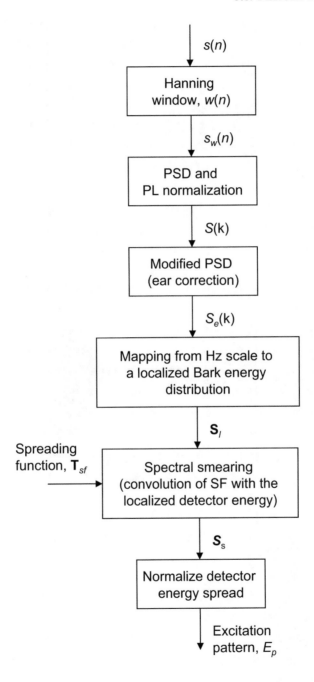

Figure 3.25: A flowchart illustrating the AEP computation using the parametric spreading function.

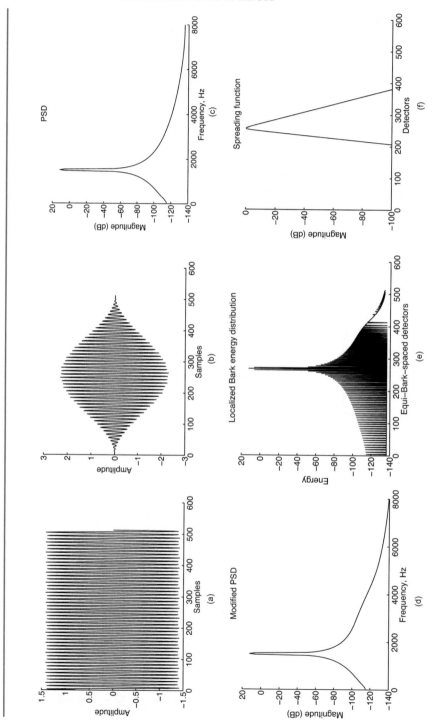

Figure 3.26: AEP computation using parametric spreading function. (a) Test signal, $s(n)$, a 1.5 kHz sinusoid, (b) Hanning windowed signal, $s_w(n)$, (c) the FFT magnitude spectrum, $s_w(n)$, (d) modified PSD after ear correction, $S_e(k)$, (e) mapping from the Hz scale to localized Bark energy distribution, $\mathbf{S}_l$. The x-axis is an equi-Bark-spaced set of 512 discrete detectors, (f) spreading function, $\mathbf{T}_{sf}$, (g) spectral smearing performed by convolving $\mathbf{S}_l$ with the spreading function, $\mathbf{T}_{sf}$, (h) excitation pattern, E_p. *Continues.*

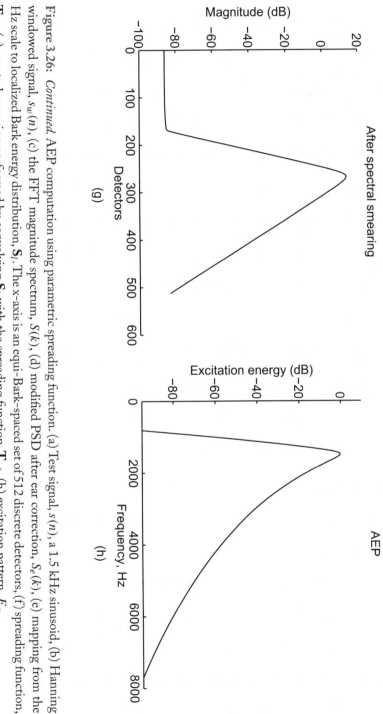

Figure 3.26: *Continued.* AEP computation using parametric spreading function. (a) Test signal, $s(n)$, a 1.5 kHz sinusoid, (b) Hanning windowed signal, $s_w(n)$, (c) the FFT magnitude spectrum, $S(k)$, (d) modified PSD after ear correction, $S_e(k)$, (e) mapping from the Hz scale to localized Bark energy distribution, $\mathbf{S}_l$. The x-axis is an equi-Bark-spaced set of 512 discrete detectors, (f) spreading function, $\mathbf{T}_{sf}$, (g) spectral smearing performed by convolving $\mathbf{S}_l$ with the spreading function, $\mathbf{T}_{sf}$, (h) excitation pattern, E_p.

```
%----------------------------------------------------------------------------
% Auditory Excitation Pattern estimation using parametric spreading function method
%----------------------------------------------------------------------------
function [AEP] = Excitation_Patt(s, fs, fft_size)

  % Freq. bins in Hz
  freq_hz = [1:fft_size/2+1] * (fs/fft_size);

  % Initialize soundfield-to-eardrum
  earComp = 6.5.*exp(-0.6*(((freq_hz/1000)-3.3).^2)) - 1e-3.*((freq_hz/1000).^4);

  % Initialize spreading function
  SF = computeSF(fs);

  % Generate mapping vector from Hz to localized Basilar energy distribution
  W = computeBasilarMapping(fft_size, fs);

  % Spectral analysis and SPL Normalization
  P = spectralAnalysis(s, fft_size);
  % Apply soundfield-to-eardrum and middle ear corrections
  Pc = P + earComp;
  % Return to linear scale
  Pc = 10.^(Pc./10);

  % Transform DFT components into a localized Basilar energy distribution
  PLoc = Pc * W;

  % Convolve localized Basilar energy distribution with the prototype spreading function
  Pspread = SpreadBasilar(PLoc.', SF);

  % Normalize spread version
  Gs = sum(Pspread)/sum(PLoc);
  Pspread = (1/Gs) * Pspread;

  % Auditory excitation pattern
  AEP = 10*log10(Pspread);

%----------------------------------------------------------------------
% Spreading function
%----------------------------------------------------------------------
function SF = computeSF(fs)

  l = 27;           % Lower skirt slope, dB per Bark
  u = 10;           % Upper skirt slope, dB per Bark
  c = 0.16;         % Corrects shift such that peak is at 0 Barks
  t = 0.1;          % Creates desired smoothness (rounded peak)
  thresh = -110;    % Threshold below which spreading weights discarded (dB)
```

Figure 3.27: AEP estimation using parametric spreading function method. *Continues.*

```
% Lower bound on frequencies of interest
zmin = hztoBark(20);
% Upper frequency bound
zmax = hztoBark(fs/2);
% Number of detectors used between fmin and fmax
D = 512;
% Detector resolution, in Barks
res = (zmax - zmin)/(D-1);

% Initialize the spreading function
SF=zeros(2*D-1,1);
x=((1-D)*res):res:((D-1)*res);
a1=(l-u)/2;
a2=x+c;
a3=(l+u)/2;
a4=sqrt(t+(x+c).^2);
SF=a1.*a2-a3.*a4;
% Normalize to 0 dB
SF=SF-(max(SF));
SF(find(SF<thresh))=thresh;
SF=10.^(0.1*SF);
SF=SF(:);

% --------------------------------------------------------------------------
% Generate mapping matrix from Hz to localized Basilar energy distribution
% --------------------------------------------------------------------------
function [W] = computeBasilarMapping(fft_size, fs)

    % Lower bound on frequencies of interest
    fmin = 20;
    zmin = hztoBark(fmin);
    % Upper frequency bound
    fmax = fs/2;
    zmax = hztoBark(fmax);
    % Number of detectors used between fmin and fmax
    D = 512;
    % Detector resolution, in Barks
    res = (zmax - zmin)/(D-1);
    % CB rate associated with each detector
    z = ((0:(D-1))*res).' + zmin;

    % Number of spectral points upto Nyquist
    M = fft_size/2+1;
    % DFT filter bank frequency step
    fc = fs/fft_size;
    % Bark frequency step
    zc = hztoBark((1:M)*fc);
```

Figure 3.27: *Continued.* AEP estimation using parametric spreading function method. *Continues.*

```
% Initialize weight matrix to handle many DFT bins mapped to one detector
W = sparse(M,D);

minbin = max(1, round(fmin*fft_size/fs));
maxbin = round(fmax*fft_size/fs);
% For DFT bins, K = [minbin to maxbin] estimate the Basilar energy distribution
for K = minbin : maxbin
  [distance, detector] = min(abs(zc(K)-z));
  W(K, detector) = 1;
end


% -------------------------------------------------------------------------
% Spectral analysis
% -------------------------------------------------------------------------
function [PSD] = spectralAnalysis(s, fft_size)

  % Power normalization term, PN
  PN = 90.3;

  % Normalize the input audio samples according to the FFT length
  %  and the number of bits per sample
  x = s/fft_size;

  %  Design the Hanning window, w(n)
  win = hanning(fft_size).';

  %  Compute the power spectral density (PSD), P
  P = PN + 10*log10( (abs(fft(win.*x, fft_size))).^2 );
  PSD = P(1: fft_size/2+1);   % Only first half is required


%-------------------------------------------------------------------------
% Spread of Basilar energy distribution
%-------------------------------------------------------------------------
function basilarSpread = SpreadBasilar(basilarEnerDistr, SF)

  Dh = 900;  % Select subset of spreading function (upper skirt)
  Dl = 400;  % lower skirt
  D = 512;   % Number of detectors

  % Initialize the spread of Basilar energy distribution
  basilarSpread = zeros(D,1);
  for iD = 1:D
    x1 = min(iD-1, Dl);
    x2 = min(D-iD, Dh);
    k = (iD-x1):(iD+x2);
    basilarSpread(iD) = basilarEnerDistr(k)'*SF(iD-k+D);
  end
```

Figure 3.27: *Continued.* AEP estimation using parametric spreading function method. *Continues.*

```
function z = hztoBark(f)
  fKhz = f/1000;
  z = 13 * atan(0.76*fKhz) + 3.5 * atan((fKhz/7.5).^2);

%%% Bark to Hz conversion
function f = barktoHz(z)
  % Use interpolated table lookup
  fi = [20 : 20000];
  zi = Hz2Bark(fi);
  f = interp1(zi,fi,z,'spline');
```

Figure 3.27: *Continued.* AEP estimation using parametric spreading function method.

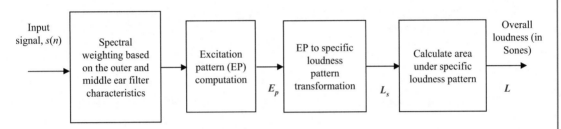

Figure 3.28: Perceptual loudness computation.

First, using the parametric spreading function method described in Section 3.5.2, the excitation, E_p, is computed. Second, the AEP is transformed to specific loudness pattern, L_s, as follows.

$$L_s = kE_p^\alpha \tag{3.31}$$

where k and α are taken as 0.047 and 0.3, respectively. Note that the above equation is a simplified form of the loudness model, $L_s = k\left[(GE_p + A)^\alpha - A^\alpha\right]$, in that the effects of the low sound levels ($A = 0$) are not considered and the gain, G, associated with the cochlear amplifier at low frequencies is set to one. The overall loudness of a sound is obtained by summing the specific loudness, i.e., the loudness per equi-spaced-Bark detector, across the whole Bark scale. In particular, the total area under the specific loudness pattern gives us the overall loudness, L, in sones,

$$L = \int L_s(z_b)dz_b . \tag{3.32}$$

Figure 3.29 (a) through (d) show the plots of the input signal, modified PSD, the AEP, and the specific loudness pattern. The overall loudness level is 103 phons and the loudness is 90.7 sones. Note that the sone is a unit of perceived loudness. One sone is equivalent to 40 phons, which is defined as the loudness of a 1-kHz tone at 40 dB SPL. The number of sones to a phon was chosen such that a doubling of the number of sones sounds like a doubling of the loudness. For example,

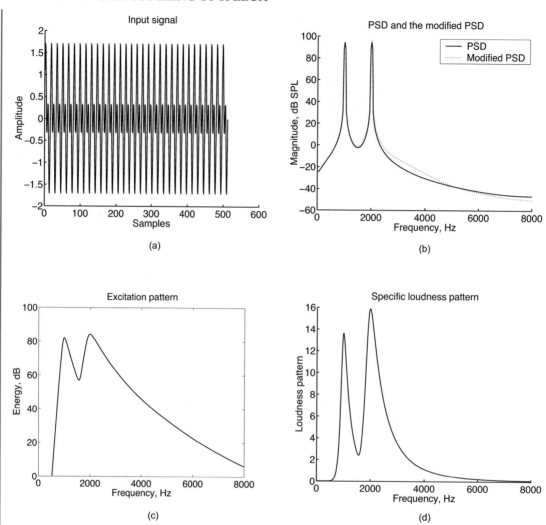

Figure 3.29: An example depicting the loudness computation. (a) input signal, (b) the PSD and the modified PSD, (c) the AEP, and (d) the specific loudness pattern.

50 phons equals 2 sones, 60 phons equals 4 sones, 70 phons equals 8 sones, 80 phons equals 16 sones, 90 phons equals 32 sones, and so on.

3.7 SUMMARY

A review of various auditory representations is given in this Chapter. In particular, we studied in detail the auditory excitation pattern (AEP) and the perceptual loudness. A signal processing perspective of the human auditory system was included. Both simultaneous and temporal masking concepts were presented. Masking asymmetry was explained with an example. The ISO/IEC MPEG psychoacoustic models 1 and 2 were described in detail using flowcharts. A step-by-step procedure to compute AEPs and the PL was described.

Bibliography

[1] M. R. Schroeder, B. S. Atal, and J. L. Hall, "Optimizing digital speech coders by exploiting masking properties of the human ear," *J. Acoust. Soc. Amer.*, vol. 66, no. 6, p. 1647, 1979. DOI: 10.1121/1.383662 1, 3, 55

[2] B. S. Atal and M. R. Schroeder, "Predictive coding of speech signals and subjective error criteria," *IEEE Trans. Acoust., Speech, Signal Processing*, vol. 27, no. 3, pp. 247–253, June 1979. DOI: 10.1109/TASSP.1979.1163237 1, 4, 10, 53

[3] M. R. Schroeder and B. S. Atal, "Code-excited linear prediction (CELP): High quality speech at very low bit rates," in *Proc. IEEE Int. Conf. Acoust., Speech Signal Processing*, vol. 10, Apr. 1985, pp. 937–940. DOI: 10.1109/ICASSP.1985.1168147 1, 3

[4] B. S. Atal and M. R. Schroeder, "Stochastic coding of speech signals at very low bit rates," in *Proc. Int. Conf. Comm.*, May 1984, pp. 1610–1613. DOI: 10.1016/0167-6393(85)90043-3 1, 3

[5] ITU, *Coding of Speech at 8kb/s using Conjugate-Structure Algebraic-Code-Excited Linear- Prediction* (CS-ACELP), ITU Study Group 15 Std. Draft Recommendation G.729, 1995. 1, 4, 6, 7, 50

[6] TIA, *Selectable Mode Vocoder* (SMV) *Service Option for Wideband Spread Spectrum Communications Systems*, TIA Std. TIA IS-893, 2004. 1, 6, 7

[7] TIA, *Enhanced variable rate codec, Speech service option 3 for wideband spread spectrum digital systems*, TIA Std. TIA IS-127, 1997. 1, 6, 7

[8] S. F. Boll, "Suppression of acoustic noise in speech using spectral subtraction," *IEEE Trans. Acoust., Speech, Signal Processing*, vol. 27, no. 2, pp. 113–120, Apr. 1979. DOI: 10.1109/TASSP.1979.1163209

[9] Y. Ephraim and D. Malah, "Speech enhancement using a minimum mean-square error short-time spectral amplitude estimation," *IEEE Trans. Acoust., Speech, Signal Processing*, vol. 32, no. 6, pp. 1109–1121, Dec. 1984. DOI: 10.1109/TASSP.1984.1164453

[10] Y. Ephraim and D. Malah, "Speech enhancement using a minimum mean-square error log-spectral amplitude estimator," *IEEE Trans. Acoust., Speech, Signal Processing*, vol. 33, no. 2, pp. 443–445, Apr. 1985. DOI: 10.1109/TASSP.1985.1164550 1

100 BIBLIOGRAPHY

[11] J. D. Markel and A. H. Gray, *Linear prediction of speech*. Springer, 1976. 2, 9, 10, 13, 16, 30

[12] B. S. Atal and S. L. Hanauer, "Speech analysis and synthesis by linear prediction of the speech wave," *J. Acoust. Soc. Amer.*, vol. 50, no. 2, pp. 637–655, 1971. DOI: 10.1121/1.1912679 2, 10, 15, 16

[13] D. Florencio, "Investigating the use of asymmetric windows in CELP vocoders," in *Proc. IEEE Int. Conf. Acoust., Speech Signal Processing*, vol. 2, Apr. 1993, pp. 427–430. DOI: 10.1109/ICASSP.1993.319331 2

[14] F. Itakura, "Line spectrum representation of linear predictive coefficients," *J. Acoust. Soc. Amer.*, vol. 57, no. 1, p. S35, 1975. DOI: 10.1121/1.1995189 3, 10

[15] K. K. Paliwal and B. S. Atal, "Efficient vector quantization of LPC parameters at 24 bits/frame," *IEEE Trans. Speech Audio Processing*, vol. 1, no. 1, pp. 3–14, Jan. 1992. DOI: 10.1109/89.221363 3, 6, 10, 16, 49

[16] P. Kroon and W. B. Kleijn, "Linear prediction-based analysis-synthesis coding," in *Speech coding and synthesis*, W. B. Kleijn and K. K. Paliwal, Eds. Amsterdam, The Netherlands: Elsevier Science, 1995, ch. 3, pp. 79–119. 4, 16, 19

[17] R. Salami *et al.*, "Design and description of CS-ACELP: A toll quality 8 kb/s speech coder," *IEEE Trans. Speech Audio Processing*, vol. 6, no. 2, pp. 116–130, Mar. 1998. DOI: 10.1109/89.661471 3, 4, 5, 6, 7, 13

[18] B. S. Atal and J. Remde, "A new model for LPC excitation for producing natural sounding speech at low bit rates," in *Proc. IEEE Int. Conf. Acoust., Speech Signal Processing*, vol. 7, Apr. 1982, pp. 614–617. DOI: 10.1109/ICASSP.1982.1171649 3

[19] S. Singhal and B. S. Atal, "Improving performance of multi-pulse LPC coders at low bit rates," in *Proc. IEEE Int. Conf. Acoust., Speech Signal Processing*, vol. 9, Mar. 1984, pp. 9–12.

[20] P. Kroon and B. S. Atal, "Predictive coding of speech using analysis-by-synthesis techniques," in *Twenty-Fourth Asilomar Signals, Systems and Computers Conference*, vol. 2, Nov. 1990, pp. 664–668. DOI: 10.1109/ACSSC.1990.523421 3

[21] ISO, *Information Technology - Coding of Moving Pictures and Associated Audio for Digital Storage Media at up to About 1.5 Mb/s, Part 3: Audio*, ISO/IEC JTC1/SC29/WG11 MPEG Std. IS 11 172–3, 1992. 3, 53, 54, 64

[22] ISO, *Information Technology - Generic Coding of Moving Pictures and Associated Audio, Part 3: Audio*, ISO/IEC JTC1/SC29/WG11 MPEG Std. IS13 818–3, 1994.

[23] ISO, *Generic Coding of Moving Pictures and Associated Audio: Audio (non backwards compatible coding, MPEG-2 NBC/AAC)*, ISO/IEC JTC1/SC29/WG11 MPEG Std. Committee Draft 13 818–7, 1996. 3

[24] G. Davidson, "Digital audio coding: Dolby AC-3," in *The Digital Signal Processing Handbook*, V. Madisetti and D. Williams, Eds. CRC Press, 1998, ch. 41, pp. 1–21. 3, 53

[25] M. Davis, "The AC-3 multichannel coder," in *Proc. 95th Conv. Aud. Eng. Soc.*, no. 3774, 1993. 53

[26] S. Smyth, W. Smith, M. Smyth, M. Yan, and T. Jung, "DTS coherent acoustics delivering high quality multichannel sound to the consumer," in *Proc. 100th Conv. Aud. Eng. Soc.*, no. 4293, 1996. 3, 53

[27] Y. Medan, E. Yair, and D. Chazan, "Super resolution pitch determination of speech signals," *IEEE Trans. Signal Processing*, vol. 39, no. 1, pp. 40–48, Jan. 1991. DOI: 10.1109/78.80763 4

[28] W. C. Chu, *Speech Coding Algorithms*. New York: Wiley-Interscience, 2003. 4

[29] ITU, *Dual Rate Speech Coder for Multimedia Communications transmitting at 5.3 and 6.3 kb/s*, ITU Std. Draft Recommendation G.723.1, 1995. 4, 6, 7

[30] A. Kataoka, T. Moriya, and S. Hayashi, "An 8-kb/s conjugate structure CELP (CS-ACELP) speech coder," *IEEE Trans. Speech Audio Processing*, vol. 4, no. 6, pp. 401–411, Nov. 1993. DOI: 10.1109/89.544525 4, 6

[31] B. S. Atal, "Predictive coding of speech at low bit rates," *IEEE Trans. Commun.*, vol. 30, no. 4, pp. 600–614, Apr. 1982. DOI: 10.1109/TCOM.1982.1095501 4

[32] J. P. Adoul *et al.*, "Fast CELP coding based on algebraic codes," in *Proc. IEEE Int. Conf. Acoust., Speech Signal Processing*, vol. 12, Apr. 1987, pp. 1957–1960. DOI: 10.1109/ICASSP.1987.1169413 4, 6

[33] C. La°amme *et al.*, "On reducing computational complexity of codebook search in CELP coder through the use of algebraic codes," in *Proc. IEEE Int. Conf. Acoust., Speech Signal Processing*, vol. 1, Apr. 1990, pp. 177–180. DOI: 10.1109/ICASSP.1990.115567 4

[34] J.-H. Chen and A. Gersho, "Adaptive post-filtering for quality enhancement of coded speech," *IEEE Trans. Speech Audio Processing*, vol. 3, no. 1, pp. 59–70, Jan. 1995. DOI: 10.1109/89.365380 5, 6

[35] V. Atti and A. Spanias, "Speech analysis by estimating perceptually relevant pole locations," in *Proc. IEEE Int. Conf. Acoust., Speech Signal Processing*, vol. 1, Philadelphia, PA, Mar. 2005, pp. 217–220. DOI: 10.1109/ICASSP.2005.1415089

[36] V. Atti and A. Spanias, "Rate determination based on perceptual loudness," in *Proc. IEEE Int. Symp. Circuits and Systems*, vol. 2, May 2005, pp. 848–851. DOI: 10.1109/ISCAS.2005.1464721

[37] B. Wei and J. Gibson, "A new discrete spectral modeling method and an application to CELP coding," *IEEE Signal Processing Lett.*, vol. 10, no. 4, pp. 101–103, Apr. 2003. DOI: 10.1109/LSP.2003.808550 10, 34, 35

[38] D. Petrinovic, "Discrete weighted mean square all-pole modeling," in *Proc. IEEE Int. Conf. Acoust., Speech Signal Processing*, vol. 1, Apr. 2003, pp. 828–831. DOI: 10.1109/ICASSP.2003.1198909 10, 34, 35

[39] P. Alku and T. Backstrom, "Linear predictive method for improved spectral modeling of lower frequencies of speech with small prediction orders," *IEEE Trans. Speech Audio Processing*, vol. 12, no. 2, pp. 93–99, Mar. 2004. DOI: 10.1109/TSA.2003.822625 10, 11, 41

[40] H. Hermansky, "Perceptual linear predictive (PLP) analysis of speech," *J. Acoust. Soc. Amer.*, vol. 87, no. 4, pp. 1738–1752, Apr. 1990. DOI: 10.1121/1.399423 10, 14, 19, 20, 21

[41] H. W. Strube, "Linear prediction on a warped frequency scale," *J. Acoust. Soc. Amer.*, vol. 68, no. 4, pp. 1071–1076, Oct. 1980. DOI: 10.1121/1.384992 10, 14, 21

[42] A. El-Jaroudi and J. Makhoul, "Discrete all-pole modeling," *IEEE Trans. Signal Processing*, vol. 39, no. 2, pp. 411–423, Feb. 1991. DOI: 10.1109/78.80824 10, 30, 33, 34, 35

[43] E. Zwicker and H. Fastl, *Psychoacoustics Facts and Models*. Springer-Verlag, 1990. 14, 18, 53, 58, 84, 85

[44] B. C. J. Moore, *An Introduction to the Psychology of Hearing*. Academic Press, 2003. 14, 18, 53, 58, 84

[45] B. Bessette *et al.*, "The adaptive multirate wideband speech codec (AMR-WB)," *IEEE Trans. Speech Audio Processing*, vol. 10, no. 8, pp. 620–636, Nov. 2002. DOI: 10.1109/TSA.2002.804299 6, 7, 8, 10, 49, 55

[46] A. Harma and U. K. Laine, "A comparison of warped and conventional linear prediction," *IEEE Trans. Speech Audio Processing*, vol. 9, no. 5, pp. 579–588, July 2001. DOI: 10.1109/89.928922 10, 21, 29, 30

[47] T. Painter and A. Spanias, "Perceptual coding of digital audio," *Proc. of the IEEE*, vol. 88, no. 4, pp. 451–513, Apr. 2000. DOI: 10.1109/5.842996 61, 63

[48] J. Makhoul, "Linear prediction: A tutorial review," *Proc. of the IEEE*, vol. 63, no. 5, pp. 561–580, Apr. 1975. DOI: 10.1109/PROC.1975.9792 9, 10, 13

[49] J. Makhoul, "Spectral linear prediction: properties and applications," *IEEE Trans. Acoust., Speech, Signal Processing*, vol. 23, no. 3, pp. 283–296, June 1975. DOI: 10.1109/TASSP.1975.1162685 10, 30

[50] R. J. McAulay, "Maximum likelihood spectral estimation and its application to narrow-band speech coding," *IEEE Trans. Acoust., Speech, Signal Processing*, vol. 32, no. 2, pp. 243–251, Apr. 1984. DOI: 10.1109/TASSP.1984.1164318

[51] C. Lee, "Robust linear prediction for speech analysis," in *Proc. IEEE Int. Conf. Acoust., Speech Signal Processing*, vol. 12, Apr. 1987, pp. 289–292. DOI: 10.1109/ICASSP.1987.1169680

[52] J. Picone, "Joint estimation of the LPC parameters and the multipulse excitation," *Speech Commun.*, vol. 5, no. 3/4, pp. 253–260, Dec. 1986. DOI: 10.1016/0167-6393(86)90012-9

[53] R. Mizoguchi, "Speech analysis by selective linear prediction in the time domain," in *Proc. IEEE Int. Conf. Acoust., Speech Signal Processing*, vol. 3, May 1982, pp. 1573–1576. DOI: 10.1109/ICASSP.1982.1171428

[54] F. Itakura and S. Saito, "An analysis-synthesis telephony based on maximum likelihood method," in *Proc. 6th Int. Conf. Acoustics*, 1968, pp. 17–20. 10

[55] A. Gray and J. Markel, "Distance measures for speech processing," *IEEE Trans. Acoust., Speech, Signal Processing*, vol. 24, no. 5, pp. 380–391, Oct. 1976. DOI: 10.1109/TASSP.1976.1162849 30

[56] TIA, *Speech service option standard for wideband spread spectrum digital cellular system*, TIA Std. TIA IS-96, 1996. 6, 7

[57] A. Das, E. Paksoy, and A. Gersho, "Multimode and variable-rate coding of speech," in *Speech Coding and Synthesis*, W. B. Kleijn and K. K. Paliwal, Eds. Amsterdam, The Netherlands: Elsevier Science, 1995, ch. 7, pp. 257–288.

[58] A. Gersho and E. Paksoy, "Variable rate speech coding for cellular networks," in *Speech and Audio Coding for Wireless and Network Applications*, B. A. V. Cuperman and A. Gersho, Eds. Kluwer Academic Publishers, 1993, ch. 10, pp. 77–84. 7

[59] S. Ramprashad, "The multimode transform predictive coding paradigm," *IEEE Trans. Speech Audio Processing*, vol. 11, no. 2, pp. 117–129, Mar. 2003. DOI: 10.1109/TSA.2003.809195 6

[60] R. Vafin and W. B. Kleijn, "Rate-distortion optimized quantization in multistage audio coding," *IEEE Trans. Speech Audio Processing*, vol. 14, no. 1, pp. 311–320, Jan. 2006. DOI: 10.1109/TSA.2005.854104 6

[61] A. Spanias, "Speech coding: A tutorial review," *Proc. of the IEEE*, vol. 82, no. 10, pp. 1541–1582, Oct. 1994. DOI: 10.1109/5.326413

[62] Y. Bistritz and S. Pellerm, "Immittance spectral pairs (ISP) for speech encoding," in *Proc. IEEE Int. Conf. Acoust., Speech Signal Processing*, vol. 2, Apr. 1993, pp. 9–12. DOI: 10.1109/ICASSP.1993.319215 10, 49

[63] A. Harma and U. K. Laine, "Linear predictive coding with modified filter structures," *IEEE Trans. Speech Audio Processing*, vol. 9, no. 8, pp. 769–777, Nov. 2001. DOI: 10.1109/89.966080 10, 35, 36

[64] A. C. den Brinker *et al.*, "IIR-based pure linear prediction," *IEEE Trans. Speech Audio Processing*, vol. 12, no. 1, pp. 68–75, Jan. 2004. DOI: 10.1109/TSA.2003.815524 10, 11, 37

[65] T. Backstrom and P. Alku, "All-pole modeling technique based on weighted sum of LSP polynomials," *IEEE Signal Processing Lett.*, vol. 10, no. 6, pp. 180–183, 2003. DOI: 10.1109/LSP.2003.811635 11, 37, 38

[66] L. B. Jackson and S. L. Wood, "Linear prediction in cascade form," *IEEE Trans. Acoust., Speech, Signal Processing*, vol. 26, no. 6, pp. 518–528, Dec. 1978. DOI: 10.1109/TASSP.1978.1163155 11, 42, 44, 47

[67] W. C. Chu, "Window optimization in linear prediction analysis," *IEEE Trans. Speech Audio Processing*, vol. 11, no. 6, pp. 626–635, Nov. 2003. DOI: 10.1109/TSA.2003.818213 13

[68] L. Rabiner and B.-H. Juang, *Fundamentals of Speech Recognition*. Prentice Hall PTR, 1993. 18, 35, 58

[69] J. O. Smith and J. S. Abel, "Bark and ERB bilinear transforms," *IEEE Trans. Speech Audio Processing*, vol. 7, no. 6, pp. 697–708, Nov. 1999. DOI: 10.1109/89.799695 18, 20, 25, 58

[70] B. C. J. Moore and B. Glasberg, "Suggested formulae for calculating auditory-filter bandwidths and excitation patterns," *J. Acoust. Soc. Amer.*, vol. 74, pp. 750–753, 1983. DOI: 10.1121/1.389861 53, 58, 85, 87

[71] H. Hermansky and N. Morgan, "RASTA processing of speech," *IEEE Trans. Speech Audio Processing*, vol. 2, no. 4, pp. 578–589, Oct. 1994. DOI: 10.1109/89.326616 21

[72] J.-C. Junqua, H. Wakita, and H. Hermansky, "Evaluation and optimization of perceptually-based ASR front-end," *IEEE Trans. Speech Audio Processing*, vol. 1, no. 1, pp. 39–48, Jan. 1993. DOI: 10.1109/89.221366 21

[73] A. Oppenheim, D. Johnson, and K. Steiglitz, "Computation of spectra with unequal resolution using the fast Fourier transform," *Proc. of the IEEE*, vol. 59, no. 2, pp. 299–301, Feb. 1971. DOI: 10.1109/PROC.1971.8146 21

[74] E. Kruger and H. Strube, "Linear prediction on a warped frequency scale," *IEEE Trans. Acoust., Speech, Signal Processing*, vol. 36, no. 9, pp. 1529–1531, 1988. DOI: 10.1109/29.90384 21

[75] U. Laine, M. Karjalainen, and T. Altosaar, "WLP in speech and audio processing," in *Proc. IEEE Int. Conf. Acoust., Speech Signal Processing*, vol. 3, 1994, pp. 349–352. DOI: 10.1109/ICASSP.1994.390018 21

[76] A. Harma and U. K. Laine, "Warped low-delay CELP for wideband audio coding," in *Proc. AES 17th Int. Conf. High-Quality Audio Coding*, 1999, pp. 207–215. 6, 21, 29

[77] A. Harma, "Implementation of recursive filters having delay free loops," in *Proc. IEEE Int. Conf. Acoust., Speech Signal Processing*, vol. 3, May 1998, pp. 1261–1264. DOI: 10.1109/ICASSP.1998.681674 29

[78] A. Harma, "Implementation of frequency-warped recursive filters," *Signal Processing*, vol. 80, pp. 543–548, Feb. 2000. DOI: 10.1016/S0165-1684(99)00151-6 29

[79] R. Gray, A. Buzo, A. Gray, and Y. Matsuyama, "Distortion measures for speech processing," *IEEE Trans. Acoust., Speech, Signal Processing*, vol. 28, no. 4, pp. 367–376, Aug. 1980. DOI: 10.1109/TASSP.1980.1163421 30

[80] A. El-Jaroudi and J. Makhoul, "Discrete pole-zero modeling and applications," in *Proc. IEEE Int. Conf. Acoust., Speech Signal Processing*, May 1989, pp. 2162–2165. DOI: 10.1109/ICASSP.1989.266891 34, 35

[81] B. Wei and J. Gibson, "Comparison of distance measures in discrete spectral modeling," in *Proc. IEEE Dig. Sig. Proc. Workshop*, 2000. 35

[82] W. Kautz, "Transient synthesis in the time domain," in *IEEE Trans. Circuit Theory*, vol. 1, 1954, pp. 29–39. DOI: 10.1109/TCT.1954.1083588 36, 37

[83] R. King and P. Paraskevopoulos, "Digital Laguerre filters," in *J. Circuit Theory Applications*, vol. 5, 1977, pp. 81–91. DOI: 10.1002/cta.4490050108 37

[84] P. Stoica and A. Nehorai, "The poles of symmetric linear prediction models lie on the unit circle," *IEEE Trans. Acoust., Speech, Signal Processing*, vol. 34, no. 5, pp. 1419–1426, Dec. 1986. DOI: 10.1109/TASSP.1986.1164937 41, 42

[85] T. Backstrom, P. Alku, T. Paatero, and B. Kleijn, "A time-domain interpretation for the LSP decomposition," *IEEE Trans. Speech Audio Processing*, vol. 12, no. 6, pp. 554–559, Nov. 2004. DOI: 10.1109/TSA.2004.834470 42

[86] P. C. Ching and C. C. Goodyear, "Linear prediction in cascade form," *IEE Proc. pt. E*, vol. 130, 1983. DOI: 10.1109/TASSP.1978.1163155 42

[87] A. Nehorai and D. Starer, "Adaptive pole estimation," *IEEE Trans. Signal Processing*, vol. 38, no. 5, pp. 825–838, May 1990. DOI: 10.1109/29.56028 42

[88] R. Yu and C. C. Ko, "Lossless compression of digital audio using cascaded RLS-LMS prediction," *IEEE Trans. Speech Audio Processing*, vol. 11, no. 6, pp. 532–537, Nov. 2003. DOI: 10.1109/TSA.2003.818111 42

[89] G. Schuller, B. Yu, D. Huang, and B. Edler, "Perceptual audio coding using adaptive pre- and post-filters and lossless compression," *IEEE Trans. Speech Audio Processing*, vol. 10, no. 6, pp. 379–390, Sept. 2002. DOI: 10.1109/TSA.2002.803444

[90] P. Prandoni and M. Vetterli, "An FIR cascade structure for adaptive linear prediction," *IEEE Trans. Signal Processing*, vol. 46, no. 9, pp. 2566–2571, Sept. 1998. DOI: 10.1109/78.709548 42

[91] R. Zelinski and P. Noll, "Adaptive transform coding of speech signals," *IEEE Trans. Acoust., Speech, Signal Processing*, vol. 25, no. 4, pp. 299–309, Aug. 1977. DOI: 10.1109/TASSP.1977.1162974

[92] K. Rao and P. Yip, *The Discrete Cosine Transform: Algorithm, Advantages, and Applications*. Academic Press, 1990.

[93] Y. Linde, A. Buzo, and A. Gray, "An algorithm for vector quantizer design," *IEEE Trans. Commun.*, vol. 28, no. 1, pp. 84–95, Jan. 1980. DOI: 10.1109/TCOM.1980.1094577

[94] W. Chan and A. Gersho, "Constrained-storage quantization of multiple vector sources by codebook sharing," *IEEE Trans. Commun.*, vol. 39, no. 1, pp. 11–13, Jan. 1991. DOI: 10.1109/26.68269

[95] T. Moriya *et al.*, "Extension and complexity reduction of TWINVQ audio coder," in *Proc. IEEE Int. Conf. Acoust., Speech Signal Processing*, 1996. DOI: 10.1109/ICASSP.1996.543299

[96] T. Cover and J. Thomas, *Elements of Information Theory*. New York: Wiley-Interscience, 1991.

[97] ITU, *Coding at 24 and 32 kbit/s for hands-free operation in systems with low frame loss*, ITU Std. Draft Recommendation G.722.1, 1999.

[98] ITU, *7 KHz Audio Coding within 64 kbits/s*, ITU Std. Draft Recommendation G.722, 1988. 6, 7

[99] P. Mermelstein, "G.722, a new CCITT coding standard for digital transmission of wideband audio signals," *IEEE Commun. Mag.*, vol. 26, pp. 8–15, Feb. 1988. DOI: 10.1109/35.417 7

[100] R. McAulay and T. Quatieri, "Speech analysis/synthesis based on a sinusoidal representation," *IEEE Trans. Acoust., Speech, Signal Processing*, vol. 34, no. 4, pp. 744–754, Aug. 1986. DOI: 10.1109/TASSP.1986.1164910

[101] R. McAulay and T. Quatieri, "Sinusoidal coding," in *Speech Coding and Synthesis*, W. B. Kleijn and K. K. Paliwal, Eds. Amsterdam, The Netherlands: Elsevier Science, 1995, ch. 4, pp. 121–173.

[102] T. Quatieri, *Discrete-Time Speech Signal Processing: Principles and Practice*. Prentice Hall, 2001.

[103] T. Painter and A. Spanias, "Perceptual segmentation and component selection for sinusoidal representations of audio," *IEEE Trans. Speech Audio Processing*, vol. 13, no. 2, pp. 155–158, Apr. 2005. DOI: 10.1109/TSA.2004.841050 54

[104] LPC, *Telecommunications: Analog to Digital Conversion of Radio Voice by 2400 Bit/Second Linear Predictive Coding*, National Communication System - Office Technology and Standards Std. Federal Standard 1015, 1985. 7, 47

[105] J. Campbell and T. E. Tremain, "Voiced/unvoiced classification of speech with applications to the U.S. government LPC-10e algorithm," in *Proc. IEEE Int. Conf. Acoust., Speech Signal Processing*, vol. 11, Apr. 1986, pp. 473–476. DOI: 10.1109/ICASSP.1986.1169060

[106] T. E. Tremain, "The government standard linear predictive coding algorithm: LPC-10," in *Speech Technology*, Apr. 1982, pp. 40–49. 7

[107] CELP, *Telecommunications: Analog to Digital Conversion of Radio Voice By 4800 Bit/Second Code Excited Linear Prediction (CELP)*, National Communication System - Office Technology and Standards Std. Federal Standard 1016, 1991. 6, 7

[108] J. Campbell, T. E. Tremain, and V. Welch, "The proposed federal standard 1016 4800 bps voice coder: CELP," in *Speech Technology*, vol. 5, Apr. 1990, pp. 58–64. 6, 7

[109] GSM, *GSM Digital Cellular Communication Standards: Enhanced Full-Rate Transcoding*, ETSI/GSM Std. GSM 06.60, 1996. 6, 7

[110] P. Kroon, E. Deprettere, and R. Sluyter, "Regular-pulse excitation-a novel approach to effective and efficient multipulse coding of speech," *IEEE Trans. Acoust., Speech, Signal Processing*, vol. 34, no. 5, pp. 1054–1063, Oct. 1986. DOI: 10.1109/TASSP.1986.1164946 6

[111] GSM, *GSM Digital Cellular Communication Standards: Half Rate Speech; Half Rate Speech Transcoding*, ETSI/GSM Std. GSM 06.20, 1996. 7

[112] I. Gerson and M. Jasiuk, "Techniques for improving the performance of CELP-type speech coders," *IEEE J. Select. Areas Commun.*, vol. 10, no. 5, pp. 858–865, June 1992. DOI: 10.1109/49.138990

[113] ITU, *Pulse code modulation (PCM) of voice frequencies*, CCITT Std. Draft Recommendation G.711, 1988. 7

[114] ITU, *32 kb/s adaptive differential pulse code modulation (ADPCM)*, CCITT Std. Draft Recommendation G.721, 1988. 7

[115] ITU, *Wideband coding of speech at around 16 kb/s using Adaptive Multi-rate Wideband (AMR-WB)*, ITU Std. Draft Recommendation G.722.2, 2001. 6, 7, 8, 55

[116] R. Ekudden, R. Hagen, I. Johansson, and J. Svedburg, "The adaptive multi-rate speech coder," in *Proc. IEEE Workshop on Speech Coding*, 1999, pp. 117–119. DOI: 10.1109/SCFT.1999.781503

[117] ITU, *Coding of Speech at 16 kbit/s Using Low-Delay Code Excited Linear Prediction* (LD-CELP), ITU Std. Draft Recommendation G.728, 1992. 6, 7

[118] J.-H. Chen *et al.*, "A low-delay CELP coder for the CCITT 16 kb/s speech coding standard," *IEEE J. Select. Areas Commun.*, vol. 10, no. 5, pp. 830–849, June 1992. DOI: 10.1109/49.138988 7

[119] ITU, *High Level Description of TI's 4 kb/s Coder*, Texas Instruments Std. ITU-T Q21/SG16 Rapporteur meeting, AC-99–25, 1999. 7

[120] ITU, *Conexant's ITU-T 4 kbit/s deliverables*, Conexant Systems Std. ITU-T Q21/SG16 Rapporteur meeting, AC-99–20, 1999.

[121] A. McCree *et al.*, "A 4 kb/s hybrid MELP/CELP speech coding candidate for ITU standardization," in *Proc. IEEE Int. Conf. Acoust., Speech Signal Processing*, vol. 1, May 2002, pp. 629–632. DOI: 10.1109/ICASSP.2002.1005818 7

[122] TIA, *The 8 kbit/s VSELP Algorithm*, TIA Std. TIA IS-54, 1989. 6, 7, 47

[123] I. Gerson and M. Jasiuk, "Vector sum excited linear prediction (VSELP) speech coding at 8 kbits/s," in *Proc. IEEE Int. Conf. Acoust., Speech Signal Processing*, vol. 1, Apr. 1990, pp.461–464. DOI: 10.1109/ICASSP.1990.115749 7

[124] Y. Gao *et al.*, "The SMV algorithm selected by TIA and 3GPP2 for CDMA applications," in *Proc. IEEE Int. Conf. Acoust., Speech Signal Processing*, vol. 2, 2001, pp. 709–712. DOI: 10.1109/ICASSP.2001.941013 7

[125] Y. Gao *et al.*, "EX-CELP: A speech coding paradigms," in *Proc. IEEE Int. Conf. Acoust., Speech Signal Processing*, vol. 2, 2001, pp. 689–692. DOI: 10.1109/ICASSP.2001.941008

[126] A. Spanias, "Speech coding standards," in *CRC Mobile Communications Handbook*, J. Gibson, Ed. CRC, 1999.

[127] D. Knisely, S. Kumar, S. Laha, and S. Navda, "Evolution of wireless data services: IS-95 to CDMA2000," *IEEE Commun. Mag.*, vol. 36, 1998. DOI: 10.1109/35.722150 6

[128] M. El-Sayed and J. Jaffe, "A view of telecommunications network evolution," *IEEE Commun. Mag.*, vol. 40, no. 12, pp. 74–81, Dec. 2002. DOI: 10.1109/MCOM.2002.1106163 6

[129] L.R.Rabiner and R.W.Schafer, *Digital processing of speech signals*. Englewood Cliffs, NJ: Prentice-Hall, 1978.

[130] L. Deng and D. O'Shaughnessy, *Speech Processing*. CRC Press, 2003.

[131] O. Ghitza, "Auditory models and human performance in tasks related to speech coding and speech recognition," *IEEE Trans. Speech Audio Processing*, vol. 2, no. 1, pp. 115–132, Jan. 1994. DOI: 10.1109/89.260357 53

[132] B. Paillard *et al.*, "PERCEVAL: Perceptual evaluation of the quality of audio signals," *J. Acoust. Soc. Amer.*, vol. 40, no. 1, 1992. 53, 84, 88

[133] B. C. J. Moore, B. R. Glasberg, and T. Baer, "A model for the prediction of thresholds, loudness, and partial loudness," *J. Audio Eng. Soc.*, vol. 45, no. 4, pp. 224–240, Apr. 1997. 53, 57, 84, 88

[134] M. Smyth, "White paper: An overview of the coherent acoustics coding system," Digital Theater Systems (DTS), Tech. Rep., 1999.

[135] D. Sen, D. Irving, and W. Holmes, "Use of an auditory model to improve speech coders," in *Proc. IEEE Int. Conf. Acoust., Speech Signal Processing*, vol. 2, Apr. 1993, pp. 411–414. DOI: 10.1109/ICASSP.1993.319327 53

[136] J. Thyssen, W. B. Kleijn, and R. Hagen, "Using a perception-based frequency scale in waveform interpolation," in *Proc. IEEE Int. Conf. Acoust., Speech Signal Processing*, vol. 2, Apr. 1997, pp. 1595–1598. DOI: 10.1109/ICASSP.1997.596258

[137] G. Kubin and W. B. Kleijn, "On speech codig in a perceptual domain," in *Proc. IEEE Int. Conf. Acoust., Speech Signal Processing*, vol. 1, Mar. 1997, pp. 205–208. DOI: 10.1109/ICASSP.1999.758098 53

[138] S. Voran, "Observations on auditory excitation and masking patterns," in *IEEE ASSP Workshop on Applications of Signal Processing to Audio and Acoustics*, 1995. DOI: 10.1109/ASPAA.1995.482992 53

[139] H. Purnhagen, N. Meine, and B. Edler, "Sinusoidal coding using loudness-based component selection," in *Proc. IEEE Int. Conf. Acoust., Speech Signal Processing*, vol. 2, 2002, pp. 1817–1820. DOI: 10.1109/ICASSP.2002.1006118 53, 54

[140] M. Bosi and R. Goldberg, *Introduction to Digital Audio Coding and Standards*. Springer, 2002. 53

[141] A. Spanias, T. Painter, and V. Atti, *Audio Signal Processing and Coding*. New York: Wiley-Interscience, 2006. 5, 50, 53, 54

[142] J. Beerends, A. Hekstra, A. W. Rix, and M. Hollier, "Perceptual evaluation of speech quality (PESQ) the new ITU standard for end-to-end speech quality assessment part II: Psychoacoustic model," *J. Audio Eng. Soc.*, vol. 50, no. 10, pp. 765–778, Oct. 2002. 54

[143] Thiede *et al.*, "PEAQ - The ITU standard for objective measurement of perceived audio quality," *J. Audio Eng. Soc.*, vol. 48, no. 1, pp. 3–29, Jan. 2000. 54

[144] M. Goodwin, *Adaptive Signal Models: Theory, Algorithms, and Audio Applications*. Boston: Kluwer Academic Publishers, 1998. 54

[145] R. Heusdans, R. Vafin, and W. B. Kleijn, "Sinusoidal modeling using psychoacoustic adaptive matching pursuits," *IEEE Signal Processing Lett.*, vol. 9, no. 8, pp. 262–265, Aug. 2002. DOI: 10.1109/LSP.2002.802999 54

[146] Y. Hu and P. C. Loizou, "A perceptually motivated approach for speech enhancement," *IEEE Trans. Speech Audio Processing*, vol. 11, no. 5, pp. 457–465, 2003. DOI: 10.1109/TSA.2003.815936 54

[147] F. Baumgarte, "Improved audio coding using a psychoacoustic model based on a cochlear filter bank," *IEEE Trans. Speech Audio Processing*, vol. 10, no. 7, pp. 495–503, Oct. 2002. DOI: 10.1109/TSA.2002.804536 54, 84

[148] S. van de Par, A. Kohlrausch, G. Charestan, and R. Heusdens, "A new psychoacoustical masking model for audio coding applications," in *Proc. IEEE Int. Conf. Acoust., Speech Signal Processing*, vol. 2, 2002, pp. 1805–1808. DOI: 10.1109/ICASSP.2002.1006115 54, 84

[149] D. Pickard, Wikipedia: The free encyclopedia, [Online]. Available: http://en.wikipedia.org/wiki/Image:HumanEar.jpg, Tech. Rep., 2006. 56

[150] J. Deller, J. Proakis, and J. Hansen, *Discrete-Time Processing of Speech Signals*. New York: Macmillan, 1993. 58

[151] S. Stevens and J. Volkmann, "The relation of pitch to frequency," in *American Journal of Psychology*, vol. 53, no. 3, 1940, p. 329. 58

[152] J. Johnston, "Transform coding of audio signals using perceptual noise criteria," *IEEE J. Select. Areas Commun.*, vol. 6, no. 2, pp. 314–323, Feb. 1988. DOI: 10.1109/49.608 79, 82

[153] J. Johnston *et al.*, "AT&T perceptual audio coding (PAC)," in *Collected Papers on Digital Audio Bit-Rate Reduction*, N. Gilchrist and C. Grewin, Eds. Audio Engineering Society, 1996, pp. 73–81. 79

[154] F. Baumgarte, "A nonlinear psychoacoustic model applied to the ISO MPEG layer III codec," in *Proc. 99th AES Symposium*, 1995. 84

[155] R. A. Lutfi, "Additivity of simultaneous masking," *J. Acoust. Soc. Amer.*, vol. 73, no. 1, pp. 262–267, Jan. 1983. DOI: 10.1121/1.388859 84

[156] E. Ordentlich and Y. Shoham, "Low-delay code excited linear predictive coding of wideband speech at 32 kb/s," in *Proc. IEEE ICASSP-93*, vol. 2, pp. 9–12, 1993. DOI: 10.1109/ICASSP.1991.150266 6

[157] C. Laflamme, R. Salami, and J.-P Adoul, "9.6 kbit/s ACELP coding of wideband speech," in *Speech and audio coding for wireless and network applications*, Eds., B. Atal, V. Cuperman, and A. Gersho, Kluwer Academic Publishers, 1993.

[158] J.-P. Adoul and R. Lefebvre, "Wideband speech coding," in *Speech analysis and synthesis*, Eds., W. B. Kleijn and K. K. Paliwal, Kluwer Academic Publishers, 1995.

[159] N. Jayant, J. Johnston, and Y. Shoham, "Coding of wideband speech," *Speech Commun.*, vol. 11, no. 2–3, pp. 127–138, 1992. DOI: 10.1016/0167-6393(92)90007-T 6

[160] ISO/IEC JTC1/SC29/WG11 (MPEG), International Standard ISO/IEC 14496–3 AMD-1: "Coding of Audio-Visual Objects – Part 3: Audio," 2000. ("MPEG-4 version-2"). 6

[161] N.S. Jayant, V. Lawrence, and D. Prezas, "Coding of Speech and Wideband Audio," *AT&T Tech. J.*, Vol. 69(5), pp. 25–41, Sept.-Oct. 1990. 6

[162] GSM 06.10, "GSM Full-Rate Transcoding," *Technical Report Version 3.2*, ETSI/GSM, July 1989. 49

[163] I. Boyd and C. Southcott, "A Speech Codec for the Skyphone Service," *Br. Telecom Technical J.*, vol. 6(2), pp. 51–55, April 1988. DOI: 10.1007/s10550-007-0070-0 49

[164] Itu, *G.729-based embedded variable bit-rate coder: An 8-32 kbits/s scalable wideband coder bitstream interoperable with G.729*, ITU Std. Draft Recommendation, G.729.1, 2006. 6, 7, 8

Author's Biography

ATTI VENKATRAMAN

Atti Venkatraman, PhD, is a Staff Engineer at Verance Corporation. His work focuses on the development of robust and advanced watermarking algorithms for digital cinema. Prior to Verance Corporation, he was with Acoustic Technologies, where he focused on the research and development of perceptually-based algorithms for acoustic echo cancellation, noise reduction, and audio enhancement for both cellular handset and telematics solutions. He has worked extensively on integrating perceptual signal processing methods in linear prediction. He has also contributed heavily to advanced distance learning technologies involving Java.